Lecture Notes in Earth Sciences

Vol. 1: Sedimentary and Evolutionary Cycles. Edited by U. Bayer and A. Seilacher. VI, 465 pages. 1985

Vol. 2: U. Bayer, Pattern Recognition Problems in Geology and Paleontology. VII, 229 pages. 1985.

Lecture Notes in Earth Sciences

Edited by Gerald M. Friedman and Adolf Seilacher

2

Ulf Bayer

Pattern Recognition Problems in Geology and Paleontology

Springer-Verlag
Berlin Heidelberg GmbH

Author

Dr. Ulf Bayer
Institut für Geologie und Paläontologie der Universität Tübingen
Sigwartstr. 10, D-7400 Tübingen, FRG

ISBN 978-3-540-13983-6   ISBN 978-3-540-39165-4 (eBook)
DOI 10.1007/978-3-540-39165-4

2132/3140-543210

To

Dorothee

Julia and Vincent

Preface

The research on mathematical methods and computer applications in geology since 1977 was supported by the "Sonderforschungsbereich 53, Palökologie" Tübingen, directed by A. Seilacher. During the years, several "Teilprojekte" were involved: "Konstruktionsmorphologie, Fossildiagenese, Fossilvergesellschaftungen, Fossil-Lagerstätten". During the last period of the "Sonderforschungsbereich" a special project "Quantitative Methoden der Palökologie" was established: Chapters 1 to 3 serve as a final report of the scientific activities. Further information is available in the reports of the "Sonderforschungsbereich 53".

The ideas on the seismic record in chapter 4 arose during activities on Leg 71 of the DSDP-program in 1980, and I am indebted for valuable discussions to W. Güttinger, G. Dangelmayr, D. Armbruster, H. Eikenmeier of the "Institut für Informationsverarbeitung", Tübingen.

During the years, a considerable number of people was engaged somewhere in the research activities. Here I want to express my special thanks to E. Altheimer and W. Deutschle, which were active in programming problems during several years.

Tübingen   Ulf Bayer

CONTENT

1. INTRODUCTION   1

     1.1 Mathematical Geology and Algorithmization   1

     1.2 Syntax and Semantics   3

     1.3 Stability   4

2. NOISY SYSTEMS AND FOLDED MAPS   8

     2.1 Reconstruction of Sediment-Accumulation   10
         2.1.1 Accumulation Rates and Deformations of the Time-Scale   11
         2.1.2 Estimation of Original Sediment Thickness   12
         2.1.3 Underconsolidation of Sediments -- a History Effect   16

     2.2 Intraspecific Variability of Paleontological Species   19
         2.2.1 Allometric Relationships   21
         2.2.2 The 'Ontogenetic Morphospace'   23
         2.2.3 Discontinuities in the Observed Morphospace   26

     2.3 Analysis of Directional Data   29
         2.3.1 The Smoothing Error in Two Dimensions   30
         2.3.2 Stability of Local Extrema   35
         2.3.3 Approximation and Averaging of Data   40
         2.3.4 A Topological Excursus   45
         2.3.5 Densities, Folds and the Gauss Map   46

     2.4 Reconstruction of Surfaces from Scattered Data   51
         2.4.1 The Regular Grid   51
         2.4.2 Global and Local Extrapolations   54
         2.4.3 Linear Interpolation by Minimal Convex Polygons   56
         2.4.4 Stability Problems with Minimal Convex Polygons   57
         2.4.5 Continuation of a Local Approximation   62
             A) A local continuous approximation   62
             B) Continuation of a local surface approximation   65

3. NEARLY CHAOTIC BEHAVIOR ON FINITE POINT SETS   70

     3.1 Iterated Maps   72
         3.1.1 The Logistic Difference Equation   73
         3.1.2 The Numerical Approximation of a Partial Differential Equation   76
         3.1.3 Infinite Series of Caustics   79

     3.2 Chi^2-Testing of Directional Data   82

3.3 Problems with Sampling Strategies in Sedimentology   86

    3.3.1 Markov Chains in Sedimentology   86
       A) Discrete Signals   88
       B) Equal Interval Sampling   89

    3.3.2 Artificial Pattern Formation in Stratigraphic Pseudo-Time Series   92
       A) Sampling of periodic functions   92
       B) The analysis of 'bed thickness' by equal distance samples   95

3.4 Centroid Cluster Strategies -- Chaos on Finite Point Sets   98

    3.4.1 Binary Trees   99

    3.4.2 Image Concepts   100

    3.4.3 Stability Problems with Centroid Clustering   103
       Centroid cluster strategies   104
       Instabilities between clusters   106
       Instabilities within clusters   106

3.5 Tree Patterns between Chaos and Order   112

    3.5.1 Topological Properties of Open Network Patterns   115

    3.5.2 Pattern Generators for Open Networks   120
       A) Algebraic models -- prototypes of branching patterns   120
       B) A metric model -- the Honda tree   123

    3.5.3 Morphology of Branches in Honda Trees   127
       A) Length of branches   127
       B) Branching angles -- similarity and self-similarity   129
       C) Branches and bifurcations -- a quasi-continuous approximation   131

    3.5.4 Evolution of Shape   134
       A) Trees, Peano and Jordan curves   135
       B) The outline of Honda trees   138
       C) Chance and determinism   141

4. STRUCTURAL STABLE PATTERNS AND ELEMENTARY CATASTROPHES   144

4.1 Image Recognition of Three-Dimensional Objects   146

    4.1.1 The Two-Dimensional Image of Three-Dimensional Objects   147

    4.1.2 The Skeleton of Plane Figures   151

    4.1.3 Theoretical Morphology of Worm-Like Objects   153

    4.1.4 Continuous Transformations of Form   155

4.2 Surface Inversions in the Seismic Record -- the Cusp and Swallowtail Catastrophes   157

    4.2.1 Computer Simulations of Rays, Wave Fronts and Traveltime Records   160
       Linear rays   160
       Successive wave fronts   161
       Traveltime record   163

    4.2.2 Local Surface Approximation   165

    4.2.3 Linear Rays, Caustics and the Cusp Catastrophe   167

    4.2.4 Wave Fronts and the Swallowtail Catastrophe   172

    4.2.5 Wave Front Evolution and the Traveltime Record   174

    4.2.6 The Traveltime Record as a Plane Map   176

4.2.7 Singularities on the Reflector Line   179

4.2.8 Generalized Reflection Patterns in Two and Three Dimensions   183
    A) The deformed circle and the dual cusp   183
    B) Three-dimensional patterns -- the double cusp   189

4.2.9 Distributed Receivers   191

4.3 "Parallel Systems" in Geology   198

    4.3.1 Some Examples of Parallel Systems   199

    4.3.2 Similar and Parallel Folds   201

    4.3.3 Bending at Fold Hinges -- the Hyperbolic Umbilic   206

    4.3.4 Notation of Strain   210

    4.3.5 Generalized Plane Strain in Layered Media   211

4.4 SUMMARY   214

REFERENCES   217

INDEX   226

1. INTRODUCTION

Theoretical modelling and the use of mathematical methods are presently gaining in importance since progress in both geology and mathematics offers new possibilities to combine both fields. Most geological problems are inherently geometrical and morphological , and, therefore, amenable to a classification of forms from a "Gestalt point of view". Geometrical objects have to possess an inherent stability in order to preserve their essential quality under slight deformations. Otherwise, we could hardly conceive of them or describe them, and today's observation would not reproduce yesterday's result (DANGELMAYR & GÜTTINGER, 1982). This principle has become known as 'structural stability' (THOM, 1975), i.e. the persistence of a phenomenon under all allowed perturbations. Stability is also, of course, an assumption of classical Newtonian physics, which is essentially the theory of various kinds of smooth behavior (POSTON &STEWART, 1978). However, things sometimes "jump". A new species with a different morphology appears suddenly in the paleontological record (ELDREDGE & GOULD, 1972), a fault develops, a landslide moves, a computer program becomes unstable with a certain data configuration, etc. It is, surprisingly, the topological approach which permits the study of a broad range of such phenomena in a coherent manner (POSTON &STEWART, 1978; LU, 1976; STEWART, 1982). The universal singularities and bifurcation processes derived from the concept of structural stability determine the spontaneous formation of qualitatively similar spatio-temporal structures in systems of various geneses exhibiting critical behavior (DANGELMAYR & GÜTTINGER, 1982; THOM, 1975; POSTON & STEWART, 1978; GÜTTINGER & EIKEMEIER, 1979; STEWART, 1981). In addition, this return to a "geometrization of phenomena" -- after decades of algorithmization -- comes much closer to the geologist's intuitive geometric reasoning. It is the aim of this study to elucidate, by examples, how the qualitative geometrical approach allows one to classify forms and to control the behavior of complex computer algorithms.

1.1 MATHEMATICAL GEOLOGY AND ALGORITHMIZATION

The geometrical approach dominated the "mathematization" of geology until recently the computer "changed the world". As VISTELIUS (1976) summarized in his discussion of mathematical geology:

"the restoration (or 18th century) of axiomatic ideas is due to maturation of geological sciences. ... The more mature the geological ideas in the problem are, the more the mathematical tool is determined by the geological meaning of the problem. Less mature geological problems make it necessary to introduce more routine mathematical means with restricted foundation to form geology".

Here, two types of "mathematical application" occur: The mathematical attempt to "model" a specific object -- the classical method of theoretical physics which can lead to the formulation of physical laws -- and "routine mathematical means" which can mostly be translated as "statistical methods". And, case studies by computer are widely viewed as the "mathematical methods". Usually, descriptive statistical results are treated like "physical laws", a situation strongly criticized by THOM (1979). He gave the following reasons why the tool of mathematics looses its strength as one goes down the scale of sciences:

"... the first is that those sciences which do not have as efficient tools at hand as physical laws would like to be like physics and try to appear in the eyes of other people as precise as physics. Every science wants to become mathematized because it believes that way it would be put on the same footing as fundamental physics. ... The second, internal reason now works in the reverse sense: Inasmuch as a given science does not allow for precise mathematization it opens practically indefinite working possibilities to scientists in that field, because they can make models of all kinds, with approximations, statistical hypotheses, and so on, and there is practically no limit to the possibility of building models in situations which actually do not allow for specific, exact quantitative models. ... And the third reason, of course, is the computer industry's lobby: Every laboratory wants to have its own computer working even in situations where a priori there is no reason to believe that you can extract any kind of useful information out of the things you have put into the computer."

It was not Thom's aim to blame those sciences which are not as precise as physics. Rather this was directed against the degradation of the mathematical tool -- may be the reason is that topologists like Thom "want qualities -- though these sometimes acquire a fearsomely algebraic, even numerical, expression" (POSTON & STEWART, 1978). Of special interest is Thom's second argument, the indefinite working possibilities. It is always a very striking experience in applying "computer methods" that some of these methods allow for various and contradictory interpretations of the same data -- and, furthermore, that some methods can even be influenced by the ordering of the input data. Such observations were the starting point to analyze the qualitative behavior of propagated algorithms in geology.

In geology and paleontology statistical and approximation methods are generally used as strategies of pattern recognition. A density distribution is estimated from a sample, a surface is reconstructed from scattered data points, periodicity patterns of profiles are analyzed by means of statistical time series analysis, etc. Alternatively, data are sorted, grouped and classified by using factor analysis, cluster or discriminant analysis, and so forth. These are the fields where "routine mathematical methods" dominate, and it is the field where the computer allows one to analyze everything without regard to any a priori scientific meaning and without the formulation of a scientific hypothesis. During several years of work with the computer, and implementing computer programs at the 'Sonderforschungsbereich 53, Palökologie -- University Tübingen', it was a challenge to accept that rather sophisticated pattern-recognition programs may become rather unstable if some initial conditions, e.g. the input data, do not satisfy the proper conditions, and that it is, in general, not known what the "proper conditions" are. On the other hand, such computer work allowed me to collect and to analyze examples of instable procedures and problems of interpretation. A collection of such examples is presented here together with the 'qualitative' analysis of instabilities.

1.2 SYNTAX AND SEMANTICS

After decades of algorithmization in science the computer provides a valuable and indispensable tool. Much work has been invested in computer science to find rules for the verification of program correctness. The idea is to solve the programming problem

"by decomposing the overall problem into precisely specified
subproblems and then verifying that if each subproblem is solved
correctly, and if the solutions are fitted together in a speci-
fied way, then the original problem will be solved correctly"
(ALAGACIC & ARBIB, 1978).

Thus, it seems not very difficult to construct "correct" programs -- as far as the syntax is concerned (WIRTH, 1972). The other problem, however, is a semantic one: The meaning of a computer output is not defined -- no matter how correct the syntax may be -- until the meaning of the input is defined and until the input is consistent with the operations within the algorithm. In the same sense, the formulation of a program is usually not only a syntactic problem, as in most cases semantics is initially involved to some extent. The problem, however, is not restricted to computer applications in a narrow sense: It occurs whenever "formulas" are applied to data. In addition to the "correctness" of algorithms, therefore, the problem of the correct application of algorithms arises and, furthermore, the question of how to "control" the computations. These are qualitative problems because semantics itself is qualitative.

The problem and the necessity of algorithm-control in the field of geological applications will be elucidated by a collection of examples. The material is ordered in three chapters. These attempt to relate the observed instabilities with current areas of research in topological, i.e. geometrical, areas. Sometimes the examples are only weakly connected with the theoretical introduction to each chapter, as a theoretical classification is not yet available for finite point sets from which most of the examples arose. However, it will be elucidated that it is commonly a question of the viewpoint -- the question what we assume as variables and what as parameters -- if we classify a problem as a discrete or a differentiable system. Such systems are commonly accounted whenever 'stability' problems arise:

Branching solutions, i.e. bifurcations, can be detected in many classical procedures: like Chi^2-testing of directional data, surface reconstruction from scattered data points and equal distance sampling in sedimentology. The widely used centroid clustering methods turn out to provide an excellent example of chaotic behavior on finite point sets. Their statistical value is strongly questioned because they lack structural stability. Smoothing of directional data on a sphere and the classical Chi^2-test for orientation data further provide examples of a degenerated bifurcation problem. A brief discussion of iterated maps gives a connection to present areas of research.

The application of the concepts of structural stability and of catastrophe theory to reflection seismics provides a classification of structurally stable singularities in two dimensions. The analysis of image inversions in terms of the local curvature of the reflector and its depth produces a catalogue of images which allows a detailed, semiquantitative on-site survey of the traveltime record. For the geologist it can provide a framework for his qualitative structural interpretation.

The concept of structural stability also provides new insights in paleontological and evolutionary problems, e.g. the analysis of the "morphospace" of paleontological species or the "bifurcation" of species. However, such models are qualitative in the narrow sense of this term and provide rather a framework for further analyses which may terminate in models which can be tested experimentally or statistically.

Mathematical details are ignored as far as possible: The object is to convey the 'spirit' of structural stability and related fields, and its application to geological pattern recognition problems. As far as mathematics is required, it is kept to a minimal level - examples of various fields are thought to be of more interest than the mathematical theory which has been summarized in various textbooks.

1.3 STABILITY

'Pattern recognition problems' as used here, cover a wide field of 'deformations' and 'instabilities'. Various types of pattern recognition problems -- which are usually solved by computer methods -- are analyzed in terms of 'topological stability'. The term 'topological stability' or 'structural stability' means that the pattern does not drastically change under a small disturbance (ANDRONOV et al., 1966; THOM, 1974; NICOLIS & PRIGOGINE; 1977). However, pattern recognition problems may result even if the disturb-

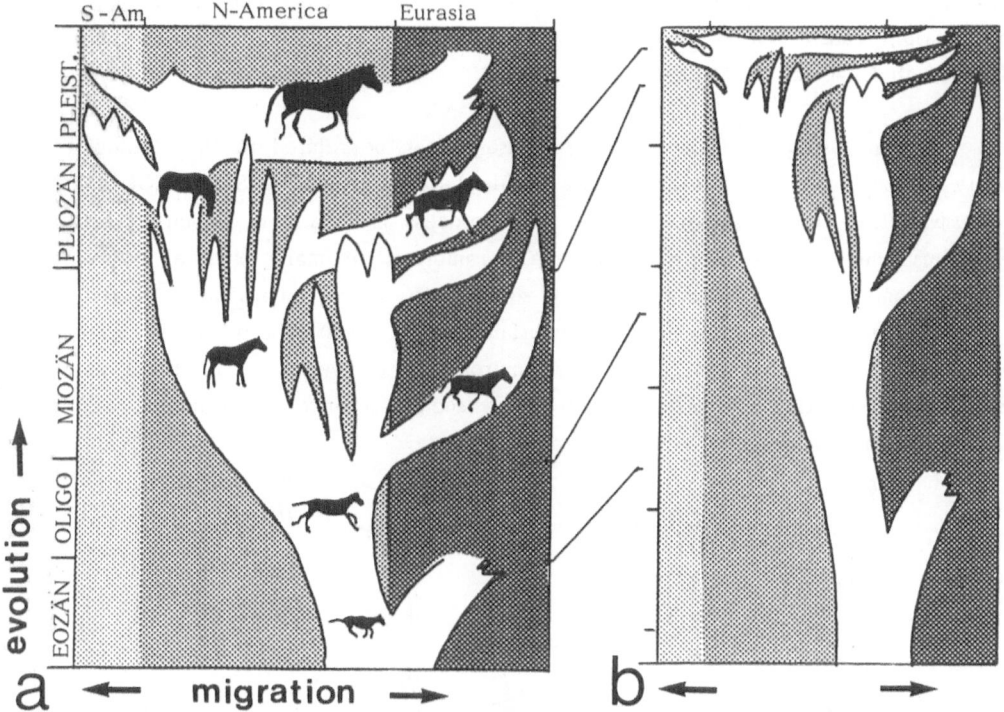

<u>Fig. 1.1:</u> The phylogenetic history of horses: (a) the classical gradual phylogeny
after SIMPSON (1951); (b) the same phylogeny redrawn along a modern time
scale. In terms of evolutionary velocities two quite different "modes of evolution" are represented by the two figures. However, in terms of phylogeny -
the relationship between species -- the transformation is structurally stable
as all pathways remain the same.

ance , the transformation, is structurally stable. Such an example is given in Fig. 1.1. The
phylogeny -- the evolutionary history -- of horses is one of the most celebrated examples
of gradual Darwinian evolution, and to some extent of directional selection (Fig. 1.1 a).
However, if the time scale used by SIMPSON (1951 and others) is replaced by the absolute
time scale under current use, the phylogenetic pattern changes dramatically with respect
to mode and velocity of evolution (Fig. 1.1 b). All significant "evolutionary events" are
now concentrated within very narrow time intervals. What does not change, is the principal structure of the phylogenetic lineages, i.e. the ancestor-descendant relationships
are structurally stable. The 'time' axis in Fig. 1.1 b is deformed like a rubber strip
which is differentially stretched without folding -- a purely topological deformation.

Although this deformation is purely topological and structurally stable, it changes
the 'semantic' interpretation of the evolutionary mode. While Fig. 1.1 a indicates a
slow gradual evolution under a long-term changing environment, Fig. 1.1 b indicates
periods of 'stasis' with little morphological evolution which are interrupted by short term

intervals of rapid evolutionary change and associated speciation events. A simple, however not trivial, transformation of the scale, thus, may transform a gradualistic picture into a punctuated one (cf. STANLEY, 1979).

Structural stability in a more precise sense can be related to the topological similarity of the "trajectories" of a process in this phase-space (NICOLIS & PRIGOGINE, 1977; HAKEN, 1977). The "internal dynamic" of a process is usually described by differential equations which usually depend on some parameters. In many physical interpretations these parameters can be identified with some state of the environment of the system, i.e. they depend on various kinds of disturbance acting continuously on the system (HAKEN, 1977). As the system and/or its environment evolves, some of these parameters can change smoothly or suddenly, and during such a change the principal behavior of the system can change.

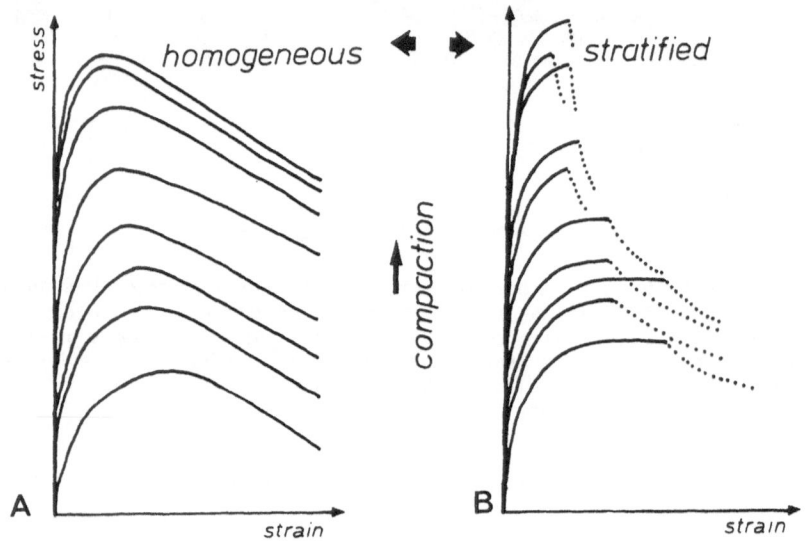

<u>Fig. 1.2:</u> Stress-strain diagrams of deep-sea sediments: a) homogeneous sediments; b) stratified sediments. Within each sequence compaction and overload increase (modified from BAYER, 1983).

In physical systems not uncommonly a threshold occurs which, when passed, causes a sudden change of the behavior of the system. The most dramatic change in a dynamic system is that its trajectories in the phase-space change their topological configuration. By a small perturbation of a parameter, the system then deviates widely from the initial situation. Fig. 1.2 illustrates this situation roughly by the "trajectories" of a stress-strain diagram. The experiments were performed with a rotating vane (with the vanes inserted parallel to the bedding planes of sediments; cf. BOYCE, 1977; BAYER, 1983). In the stress-strain diagram -- the 'phase plane' of the process -- two qualitatively very different patterns were observed (Fig. 1.2; BAYER, 1983) depending on the "stratification" of the sediments. In homogeneous sediments the stress-strain curves are smooth while in stratified sediments a sudden break occurs. Within the range of measurements these two types are independent of the compaction (preloading) of the sediments, i.e. within every set the trajectories evolve smoothly and structurally stable with increasing compaction. However, the other observed parameter, the lamination or stratification of the sediments, causes an essential change in the mode of failure: A very distinct point of failure appears in well-stratified sediments with a "sudden jump" in the stress values.

There is no common sense with regard to the term ' s t a b i l i t y ' . As HOCH-STADT (1964) notes:

> *"Often it is not necessary to determine the explicit solution to a problem, but it is important to be able to say something about the solution ... In many physical problems one is motivated by the feeling that a small change in the conditions of the problem should result in a comparably small change in the solution. ... The word stability is a very tricky word. No one definition seems to be adequate for all purposes."*

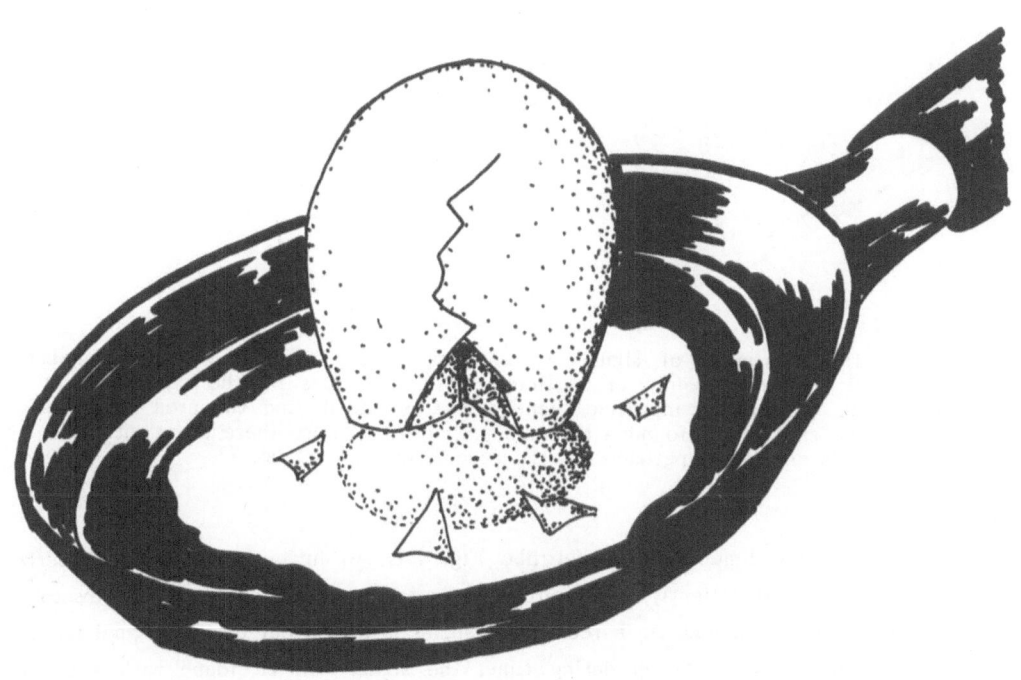

2. NOISY SYSTEMS AND FOLDED MAPS

Many geological and paleontological problems are related to the reconstruction of some ancient state from the present remains. The present state, however, is usually noisy as various factors may have influenced the system during its history. This situation comes very close to the reconstruction of deformed signals in information theory. In the theory of information the disturbance of signals by random (white) noise plays an important role (YOUNG, 1975). The stability problems of such noisy systems can be

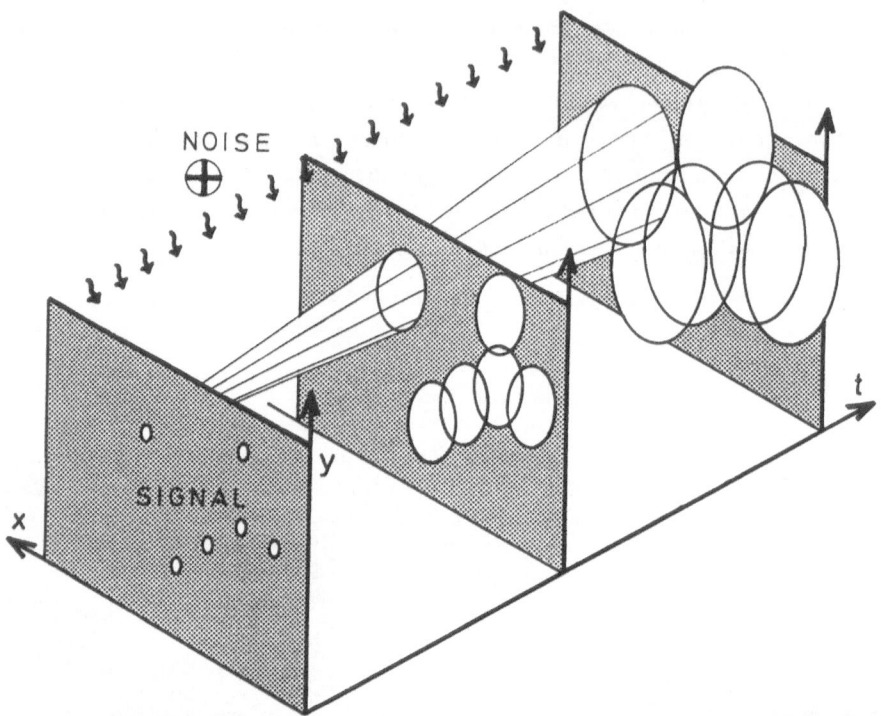

Fig. 2.1: Disturbance of signals by random, white noise. The initial signals are well separated points or sufficiently small circles on the (x,y)-plane at time t=0. With increasing time, white noise is added, and the area increases where the signals are found with some probability. Where these areas intersect, two signals are in competition for the reconstruction process.

visualized in a three-dimensional model like Fig. 2.1. An initial signal is characterized as a point (or as a sufficiently small circle) on a space plane (x,y). On its way to the receiver white noise is added. As a result, the signal is driven out of its original position. When the random noise sums up during time, the signal will be found with a specific probability within a certain area which surrounds the original position of the signal. A serious recognition problem occurs when the probabilistic neighborhoods of different signals start to overlap. Indeed, a signal found within an intersection area cannot be reconstructed with certainty.

The signals, discussed so far, are isolated points which are originally located in disjunct areas. Now, if we cover parts of the (x,y)-plane densely with signals so that their initial areas of definition are connected along boundary lines, then another way to formulate the recognition problem is more appropriate. The evolution of the signal—space along the time axis can be described as a map

source   --->   receiver.

The overlapping of the probability areas, in which a signal will be found (Fig. 2.1), can then be described as local folding of the original definition space. Fig. 2.2 illustrates the local folding of the (x,y)-plane. Within the folded areas it is not possible to solve

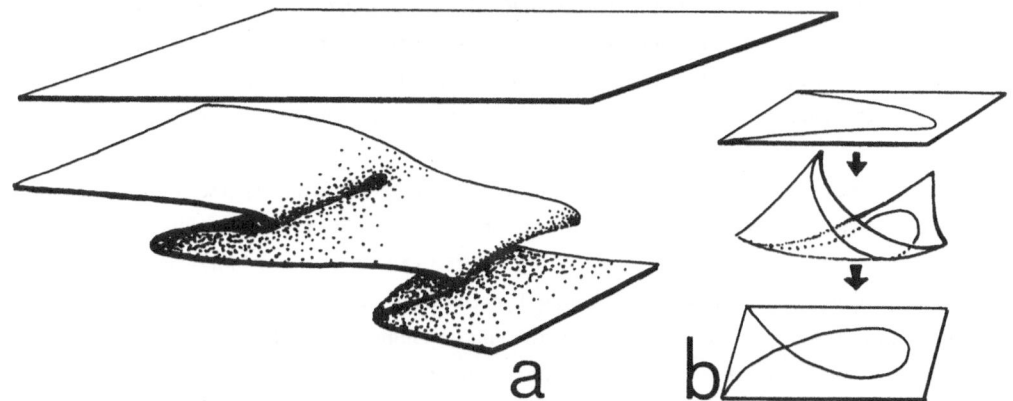

Fig. 2.2: A folded sheet as a model of a locally folded map (a). A continuous curve on a sheet may develop self-intersections in the projective plane of a folded sheet (b).

uniquely the inverse problem, the reconstruction of the 'original signal'. Fig. 2.2b illus - trates the distortion which can be caused by a local fold. A regular curve on the original plane develops a selfintersection on the projection of the local fold. However, deformations of this type will be discussed in more detail in the last chapter.

In an information-theoretic approach the disturbance of signals is due to random forces, and the reconstruction process is mainly a stochastic problem. The possible drift of signals or particles is governed by a probability distribution -- the classical example is the Brownian movement of particles in a fluid. In geology and paleontology problems of this type arise mainly if global properties of distributions are reconstructed from stochastic samples by local estimation methods. In this first chapter, pattern recognition problems will be elucidated by examples which are related to the superposition of density functions and to double- (multiple-) valued local solutions of reconstruction processes.

In the first example statistical problems are discussed in terms of the reconstruction of original sediment volumes and sediment accumulation rates. The system bears three types of disturbances: Uncertainty about the datum points; stochastic components, and a systematic trend to under-consolidation of sediments with low overburden.

Then the analysis of 'intraspecific variability of paleontological species' is discussed in terms of a probabilistic ontogenetic morphospace. It will turn out that the covariance structure between (measurable) features within the ontogenetic morphospace can be helpful for the taxonomist. But, we will see further that one cannot expect linear relationships between the features -- a nonlinear theory seems appropriate rather than just an analysis by the so-called higher statistical methods.

In a qualitative discussion of the analysis of directional data the problem will be to find an optimal weighting function for the reconstruction of a smooth density distribution from scattered data. It will become clear that the critical areas of the reconstruction are intersections of the areas, on which the weighting function is defined.

In the next example the reconstruction of surfaces from sparse point patterns by computer methods is discussed. It will be shown that the problems which arise in this context are mainly of a geometrical nature. Therefore, the algorithmization of the reconstruction process is not trivial. The local estimation methods in use provide no unique solution, i.e. they are very sensitive to small changes of the initial conditions. The example, therefore, leads over to the next chapter where this type of instabilities is discussed in more detail.

2.1 RECONSTRUCTION OF SEDIMENT-ACCUMULATION

The problem to reconstruct accumulation- and sedimentation rates arises in sedimentology in order to gather information about sea-level changes, climatic changes, and the evolution of basins. Furthermore, accumulation and sedimentation rates clearly indicate hiatuses in the sedimentary sequence on the base of which local and global events of the past are recognized (e.g. VAIL et al., 1977). However, several processes and assumptions are involved in the reconstruction of which the most important ones are

** the dating of the sediment sequence
** the estimation of the original sediment thickness without compaction.

Both reconstructions are biased and usually involve specific assumptions about the datum points, the initial porosities and the consolidation state of the sediments.

2.1.1 Accumulation Rates and Deformations of the Time-Scale

The computation of accumulation rates requires estimates of the absolute time scale, i.e. a sufficient number of datum points along the sediment column. As soon as the datum points are given, the computation of the accumulation rates is rather simple,

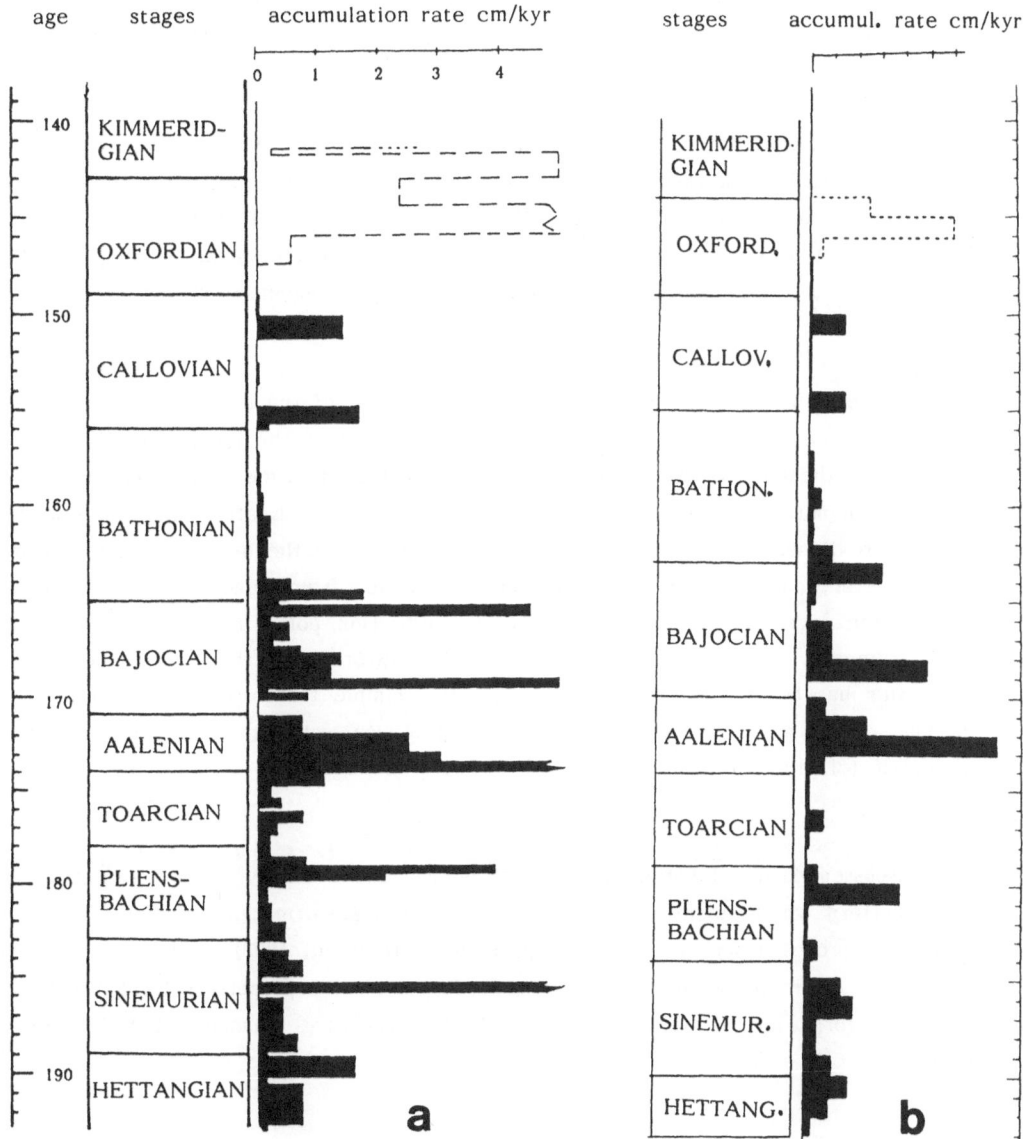

Fig. 2.3: Cyclic accumulation rates of sediments in the South German Jurassic. a) based on the time scale of VAN HINTE (1976); b) based on an altered Jurassic time scale (see text for explanation). On both scales a cyclicity is obvious, however, in (b) the cycles are much more regular (about 4 Ma), and a superimposed megatrend appears.

i.e. it is the quotient

sediment thickness
time interval of deposition .

Fig. 2.3 gives such accumulation rates for the South German Jurassic and illustrates how they subdivide the Jurassic sequence into generic depositional cycles. However, the pattern is not invariant against deformations of the time scale. Essentially the same situation arises as was discussed for the 'evolution of horses' if the time scale is altered (Fig. 2.3 A,B).

The accumulation rates in Fig. 2.3A have been calculated using the time scale of VAN HINTE (1976). However, for the time-interpolation within stages a more recent biostratigraphic subdivision (COPE et al., 1980) was used (McGHEE & BAYER, in press). A deformation of this time scale (Fig. 2.3B) alters the cyclicity drastically, and the pattern becomes much more regular. In addition, a well pronounced megacycle becomes visible (Fig. 2.3 B).

Van Hinte's Jurassic time scale is based on estimates of the upper and lower boundary of the Jurassic and of one additional 'calibration point' at the middle of the Jurassic (base of the Bathonian). Between these points he divided the time scale linearly by the number of ammonite Zones -- with the result of an average duration time of 1My for each ammonite Zone. Now, since he published his time scale, the biostratigraphic scheme has been altered, and therefore the duration time varies from Zone to Zone. However, by shifting the additional -- middle Jurassic -- calibration point into the Callovian the original assumption of 1My/ammonite Zone can be restored for the Lower and Middle Jurassic. This has been done in Fig. 2.3B, and this simple transformation generates the exceptional cyclic pattern which agrees with otherwise established cycles of similar phase length (cf. EINSELE or McGHEE & BAYER in BAYER & SEILACHER, eds., in press).

As discussed earlier for the evolution of horses, the cyclicity per se is a structural stable pattern which is preserved under topological transformations of the time scale (even if other proposed scales are used, e.g. HARLAND et al., 1978), while the regularity of the cycles, their phase length and magnitude -- i.e. all properties from which we can gather information about the velocity of the process -- change as the scale is changed.

2.1.2 Estimation of Original Sediment Thickness

One important process, which alters the physical properties of sediments, is compac-

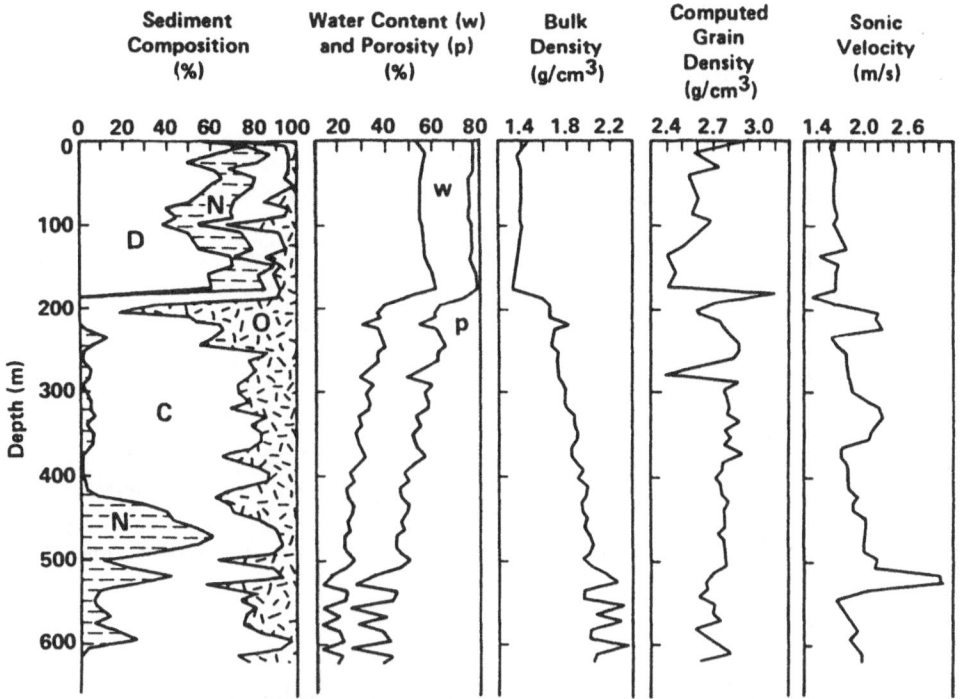

Fig. 2.4: Depth-logs for sediment composition and physical properties for DSDP-site 511 (adapted from BAYER, 1983). Data are mean values for cores (D: diatoms, N: nannofossils, c: clay content, O: other components).

tion under the overburden of later deposits. Especially in clays the physico-chemical evolution is dominated by compaction. Fig. 2.4 illustrates how the physical properties in a sediment column change with depth (i.e. overburden). In the example given, the sediment column below 200 m depth is dominated by clay, and within this column the physical parameters porosity and water-content decrease continuously as the sediment becomes increasingly compacted. In the same course the density of the sediment increases and tends slowly towards the mean grain density of the sediment.

If one assumes that the void volume of the sediments is in equilibrium with the overburden, then a first approximation for the equilibrium curve of compaction can be given by the equation

$$\frac{dn}{dp} + cn = 0 \tag{2.1}$$

(where n: porosity = relative void space; p: pressure or overburden). The constant 'c' can be interpreted as a coefficient of volume change (TERZAGHI, 1943), which is specific for particular materials. Integration gives a simple declining exponential function

$$n = n_0 \exp(-cp). \tag{2.2}$$

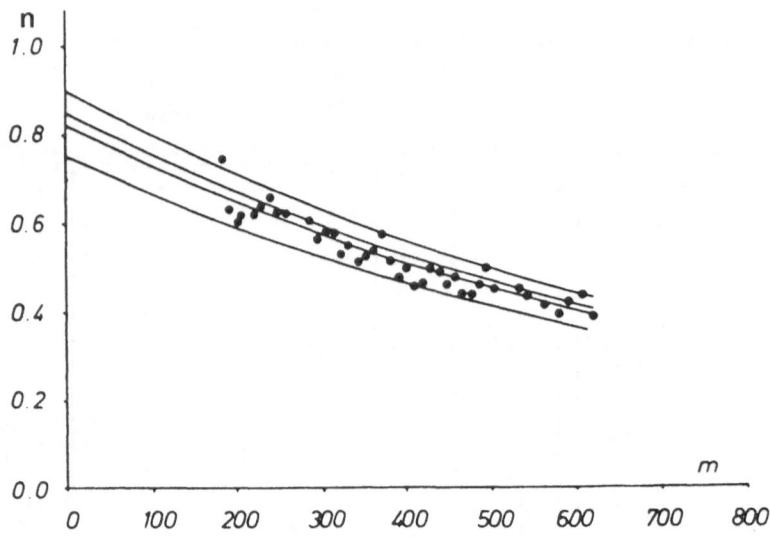

Fig. 2.5: Porosity data from Fig. 2.4 with least square fitted trend lines (see text for explanation).

This model has been used in Fig. 2.5 to estimate the decline in porosity with respect to depth (i.e. the change in bulk density has been neglected, cf. Fig. 2.4). Empirically the data are well approximated. Besides the mean trend line some more trajectories are given in Fig. 2.5, which have been constructed under the assumption that the coefficient of volume change is constant while the initial porosity of the sediments may have been variable and, thus, cause the scattering of the data points.

If we assume that the sediment is everywhere in equilibrium with the overload, it is no problem to estimate the thickness of the sediment column which would result without compaction (e.g. MAGARA, 1968; HAMILTON, 1976). Fig. 2.6 illustrates how the two principal components of a sediment change under pure compaction: The volume

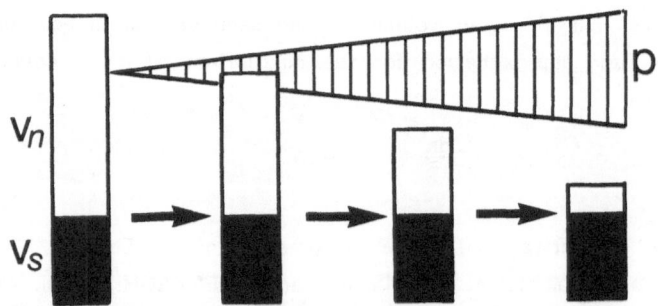

Fig. 2.6: The two components of a sediment -- voids and solids -- during compaction. Vn: volume of voids, Vs: volume of solids, p: pressure = overload.

of solids remains constant while the volume of voids decreases. The porosity is defined as the relative volume of the voids so that

$$V_n = nV \quad \text{and} \quad V_s = (1-n)V.$$

(2.3)

where n: porosity, V_n: volume of voids, and V_s: Volume of solids. Because the volume of the solids is not changed by compaction, one has

$$V_s(t=0) = V_s(t)$$

and, therefore,

$$(1-n_o)V_o = (1-n)V$$

from which we immediately have the compaction number

$$C = \frac{V}{V_o} = \frac{1-n_o}{1-n}$$

(2.4)

and the decompaction number

$$D = \frac{V_o}{V} = \frac{1-n}{1-n_o} = \frac{1}{C} .$$

(2.5)

The decompaction number allows to compute the original thickness of any sediment layer if we know its original porosity n_o, i.e.

$$V_o = DV.$$

(2.6)

The thickness of the entire sediment column then can be computed by summing up all sediment layers or, if regression curves are used as in Fig. 2.5 by evaluation of the integral

$$\sum_{i=1}^{n} \frac{1-n}{1-n_o} V_i \quad \text{or} \quad \int_{z_o}^{z} \frac{1-n(z)}{1-n_o} dz .$$

(2.7)

Both techniques have been used in Fig. 2.7 whereby the original porosities of the samples have been estimated from the intersection of their associated trajectory (from Fig. 2.5) with the zero-depth line. In a statistical sense the trajectories of Fig. 2.5 are error bounds to the mean regression line (probabilities can be attached to them by standard statistical techniques), and so the curves in Fig. 2.7 can be interpreted. Thus, the reconstruction of the original sediment amount is simply a statistical process. However, it closely resembles the situation of Fig. 2.1. The data are biased by the sampling technique as well as by the laboratory technique. Now, if we add a small error to a data point, it will not affect the results much if the overburden (or depth) is small. However, as the overburden increases, the trajectories in Fig. 2.5 come closer and closer. An error of the same magnitude, therefore, biases the results increasingly.

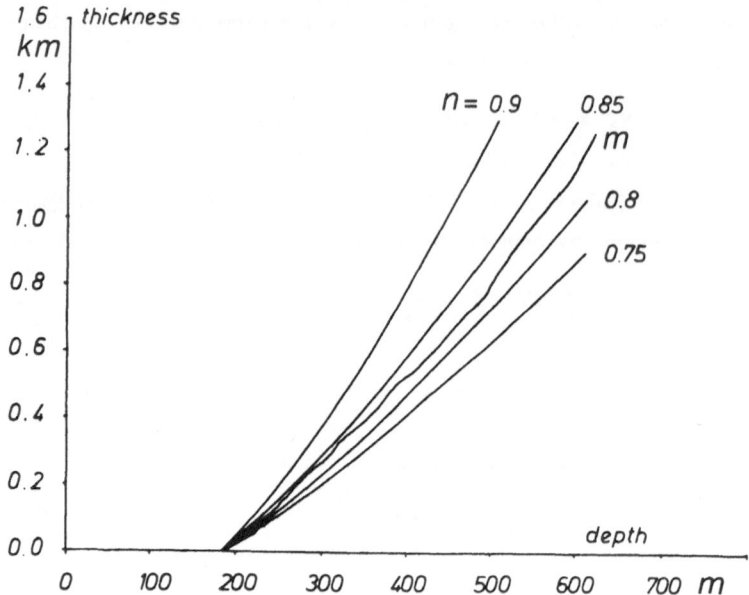

Fig. 2.7: Decompacted sediment thickness based on the data of Fig. 2.5 (m: mean trend, n=0.9 etc.: integrals of the trend lines in Fig. 2.5.

While the reconstruction of the initial state of the sediments depends on how far the "compaction machine was run", we can, on the other hand, always find the output if the "machine would work until infinity". This stable limit, of course, is simply the volume of solids, and the "dry sedimentation rates" (cf. SWIFT, 1977) are, of course, only biased by the time scale and the laboratory technique.

2.1.3 Underconsolidation of Sediments -- a History Effect

Estimations of the original sediment volume are usually based on the assumption that the consolidation state of the sediment is in equilibrium with the overburden. In this case, as was pointed out, the error of the estimation should increase with increasing overburden. However, if burial depth is small, then the time-dependent flow of the pore-water cannot be neglected; it bears on our understanding of the widespread underconsolidation of recent sediments, which is observed even under slow sedimentation rates (MAR-SAL & PHILIPP, 1970; EINSELE, 1977).

The consolidation of sediments is described by Terzaghi's model (e.g. TERZAGHI, 1943; CHILINGARIAN & WOLF, eds., 1975, 1976; DESAI & CHRISTIAN, 1977). In a one-dimensional sediment column and under the assumption of a constant coefficient of consolidation Terzaghi's model takes the form

$$\frac{\partial p}{\partial t} = m \frac{\partial^2 p}{\partial x^2}, \qquad (2.8)$$

where p: the excess pore water pressure due to overload, m: the consolidation coefficient, and t: time. If this model is discretisized in space, i.e. if the sediment column
is divided into small discrete elements, then the partial differential equation is transformed into a set of ordinary differential equations:

$$\frac{dp_x}{dt} = m(p_{x-\Delta x} - 2p_x + p_{x+\Delta x}). \qquad (2.9)$$

Now, if one reduces the system to a single element -- a situation which occurs in laboratory experiments -- then we can rewrite equation (2.9) as

$$\frac{dp}{dt} + cp = I(t), \qquad (2.10)$$

where the right side describes the "input" -- i.e. the fluxes -- at the boundaries of the
element as a function of time, and with free boundary conditions (I(t)=0) a suddenly
imposed pressure declines exponentially with time.

The idea of Terzaghi's model is that a sudden imposed load increases initially the
pore-water pressure (excess hydrostatic pressure) and that this pressure decreases after-
ward due to a loss of pore-water from the element whereby the excess hydrostatic pressure is transformed into a pressure at grain contacts. Associated with the loss of pore--
water is an increase in the number of grain contacts. Therefore, the sediment approaches
a new equilibrium state after compression which, of course, is usually not reversible.
The reduction of volume is restricted to the volume of voids, and the change in pore
volume is simply proportional to the decline in the excess hydrostatic pressure:

$$\frac{\partial n}{\partial t} dz = m \frac{\partial p}{\partial t} dz. \qquad (2.11)$$

Thus, we can solve equation (2.10) in terms of the pore volume, which in case of free
boundary conditions takes the form:

$$V(t) = V_e + (V_o - V_e) e^{-ct} \qquad (2.12)$$

for a load which is suddenly applied. The load is here represented by the equilibrium
volume V_e (cf. equation 2.2), and the excess hydrostatic pressure is proportional to the
reducible void volume $(V_o - V_e)$. Now, if at time $t=t_1$ an additional load is applied, then
equation (2.12) takes the form

$$V(t) = V_{e2} + (V(t_1) - V_{e2}) e^{-c(t-t_1)}, \qquad (2.13)$$

which can be rewritten if $V(t_1)$ is inserted from equation (2.12):

$$V(t) = V_{e2} + (V_{e1} - V_{e2})e^{ct_2} e^{-ct} + (V_o - V_{e1})e^{-ct} . \qquad (2.14)$$

As this equation shows, there is some remaining reducible porevolume from the first loading event, which has to be taken into consideration. If further load is added in discrete steps, we arrive finally at

$$V(t) = V_i + \left(\sum_{i=1}^{n} (V_{i-1} - V_i)e^{ct_i} \right) e^{-ct} \qquad (2.15)$$

which illustrates how the earlier loading states contribute to later states. The equilibrium can only be approached if the time intervals between loading are sufficiently long, otherwise the sediment layer will be underconsolidated. This history effect of loading is illustrated in Fig. 2.8 for various time intervals between loading events. The excess hydrostatic pressure (p in Fig. 3) develops clearly a maximum which degenerates to a simple declining exponential function for a single loading event and to a sequence of such single events, as the time intervals between loading become large.

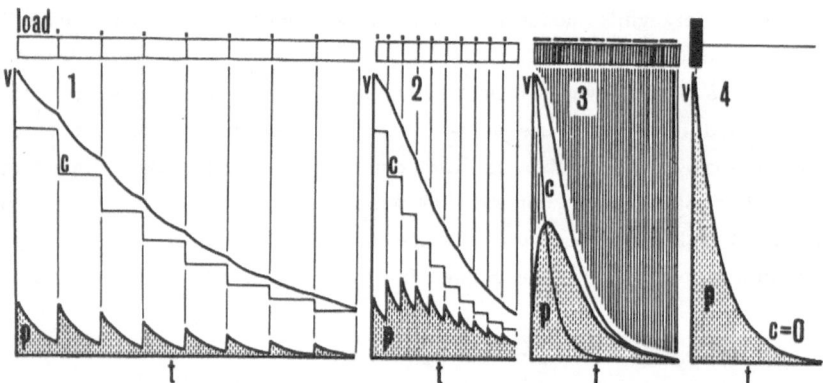

Fig. 2.8: Responce of a single sediment layer under stepwise loading when loads are applied at different time intervals: time interval decrease from left to right; right: a single load. V: void volume; P: momentary reducible void volume which will vanish even if no additional load is applied; C:equilibrium void volume for every loading event. At the top the loading intervals are marked, the total applied load is constant for all 'experiments'.

With respect to the previous discussion we have, therefore, to expect that estimates of original sediment volume are biased by the time-delays in the consolidation process, the parameters t_i in equations (2.13) and (2.14) have, of course, the structure of a time delay. Furthermore, if the pressures at the boundaries of the sediment layer are not zero, i.e. if the sediment layer is a segment within a sediment column, then the time-delay effect increases further. In case, the permeability of the sediment is low, the excess hydrostatic pressure will stay for rather long time near the values of the over-burden, and the time lack between loading and equilibrium consolidation causes a continuation of pore-water flow when sedimentation has stopped. On the other hand, if we consider a two- or three-dimensional system of strongly underconsolidated sediments, any spatial disturbance like unequal loading can initialize an instable flow of pore-water, which may lead to fluidization or liquidization of the upper sediment layers.

2.2 INTRASPECIFIC VARIABILITY OF PALEONTOLOGICAL SPECIES

In 1966, WESTERMANN observed that in several ammonite stocks -- a group of cephalopods (Fig. 2.9) -- a specific intercorrelation of morphological features occurs:

> *"Of particular interest is the inter-correlation between costation, whorl section, and coiling which has been observed in different, unrelated ammonoid stocks and cannot be satisfactorily explained"* (WESTERMANN, 1966).

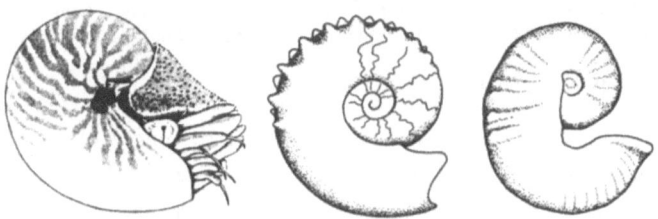

Fig. 2.9: Ectocochliate cephalopods, left recent *Nautilus* and two ammonites with well marked ontogenetic changes in morphology.

Because BUCKMAN (1892) observed, probably for the first time, this particular type of covariation (intercorrelation) between the ornament and the whorl section in ammonites, Westermann named this relationship 'B u c k m a n ' s l a w o f c o v a r i - a n c e'.   In some cases, the 'covariance' extends to other features:

> *"in general the complexity of the suture-line increases in proportion to the decrease of ornament"*.

This caused Westermann to establish 'Buckman's second law of covariance'. Proceeding in this way, any correlation between features, which cannot be satisfactorily explained would lead to a new 'law', and 'experiments' with other ammonite stocks would soon disprove the specific correlation sufficiently 'to be a law'.

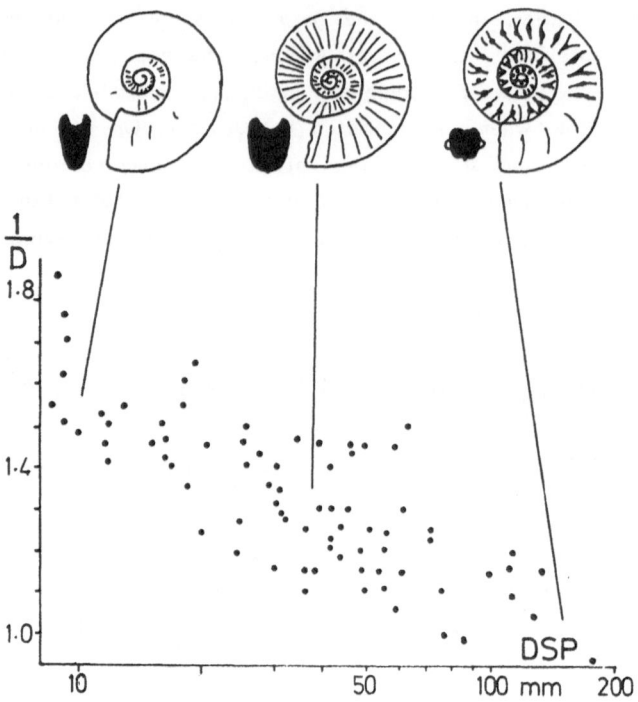

Fig. 2.10: Covariation of ornament and cross-section of *Sonninia (Euhoploceras) - adicra* (Waagen), modified from WESTERMANN (1966). The scattergram shows that the morphotypes cover a continuous area in the parameter space; D: Raup's morphological parameter "ratio of whorl height to whorl width", DSP: end diameter of the spinous stage.

On the other hand, Westermann was able to show, by means of the covariation structure, that 80 described species of the subgenus *Sonninia (Euhoploceras)* belong to a single species and that the observed variability must be viewed as an intraspecific property. His biometrical study (cf. Fig. 2.10) shows that the specimens of this lumped species fill a continuous area in the parameter space (DSP, 1/D) and that the costation types or 'forma' are regularly arranged within this parameter space (cf. Fig. 2.10 for explanation of parameters).

The covariation pattern described by 'Buckman's law' is not unique within the ammonites, but it is also not universal. Additional studies (e.g. BAYER, 1977) show that in some cases 'age effects' may play some role and that there are some special conditions which make 'Buckman's law' easily visible. One of these conditions is that the morphology changes strongly during ontogeny (cf. Fig. 2.9 for cases of rather strong ontogenetic changes). The available information makes it likely that the observed correlation is due to oblique sections through the ontogenetic morphospace because time is not accessible. The problem that age is not available in paleontology is well known; GOULD (1977) discusses in detail the problems, which arise, if equal sizes but different ages of specimens (and species) are compared by the allometric relationship.

Evidence for an age control of 'Buckman's law' comes from additional features of the shells -- the spacing of growth lines and septa -- which both are likely formed in rather regular time intervals. Especially the spacing of septa (which is more easily analyzed) shows a close relationship to cross-section and sculpture in certain ammonites (BAYER, 1972, 1977). Fig. 2.11 illustrates such a relationship between spacing of septa and shell morphology.

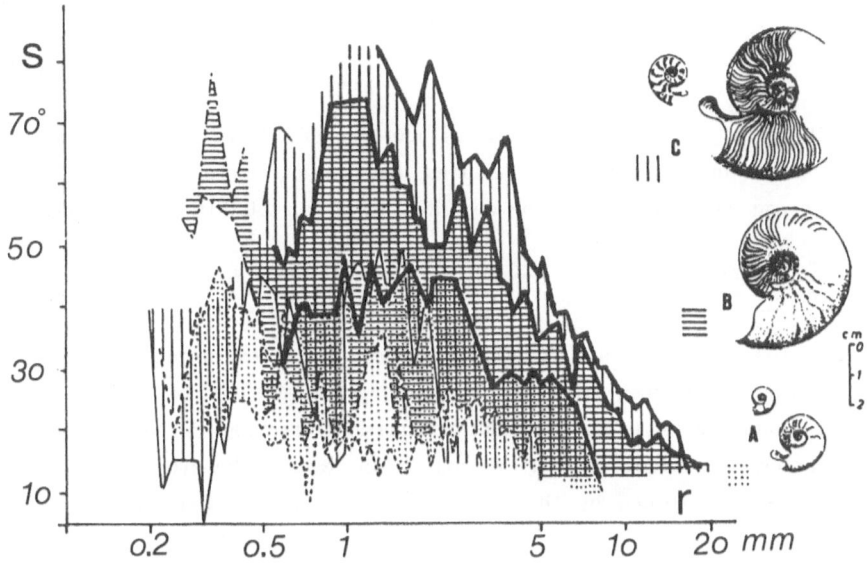

Fig. 2.11: Relationship between spacing of septa and morphology in ammonites (modified from BAYER, 1972). s: angular distance of septa; r: radius of the shell.

2.2.1 Allometric Relationships

If one accepts the hypothesis that 'Buckman's law' describes a phenomenon of intra-specific variation, we should be able to deduce it from more basic biological principles. Everyday experience on living organisms shows that most morphological features change with age and that the relationship between two morphological features (which can be quantified) leads usually to an allometric relationship, i.e. a relationship of the form:

$$y = ax^b \qquad \text{or} \qquad \log(y) = \log(a) + bx. \qquad (2.16)$$

Actually, the allometric relationship can be traced further down to the, say, 'f i r s t   p r i n c i p l e s   o f   g r o w t h' (HUXLEY, 1932). The term 'first principle' is here used in the sense that it is very likely to observe such an allometric relationship. As HUXLEY noticed, two measurements (organs etc.) are in an allometric relationship when they both grow exponentially, i.e. let y_1, y_2 be the two measurements, which grow exponentially

$$y_1 = a_1 e^{c_1 t}; \qquad y_2 = a_2 e^{c_2 t}, \qquad (2.17)$$

then by eliminating time we find the allometric relationship

$$y_1 = a_1 (\frac{y_2}{a_2})^{c_1/c_2}. \qquad (2.18)$$

Now, strictly allometric growth results also in more sophisticated growth models like the "Gompertz model". In this model one assumes that the parameter 'c' is not constant but decreases with age. Growth then can be described by a pair of differential equations

$$\frac{dy}{dt} + c(t)y = 0 \qquad \text{and an equation like} \qquad \frac{dc}{dt} = -c . \qquad (2.19)$$

The growth parameter 'c' can be any function of time, which goes to zero for large time values (ideally as time approaches infinity). Especially, any stable output of a linear control system (e.g. homogeneous linear differential equations) provides a possible input for the growth parameter. A perfect allometric relationship results whenever the two organs under consideration are controlled by the same mechanism, i.e. if their growth equations take the form:

$$\frac{dy}{dt} - c(t)*ay = 0; \qquad \frac{dx}{dt} - c(t)*by = 0; \qquad (2.20)$$

by eliminating time one finds the perfect allometric relationship

$$\frac{dy}{dx} = \frac{ay}{bx} \qquad \text{or} \qquad y = x_0 x^{a/b} .$$

In both cases considered so far the allometric relationship describes the relationship between two growing organs in the phase-plane, i.e. the trajectories of growth without consideration of the velocity of growth. Indeed, we may still further generalize the relationship to pairs of linear differential equations like

$$f(t)\frac{dy}{dt} = ax + by; \qquad f(t)\frac{dx}{dt} = cx + dy, \qquad (2.21)$$

and the relationship between the two measurements takes the form

$$\frac{dy}{dx} = \frac{ax + by}{cx + dy} \qquad (2.22)$$

23

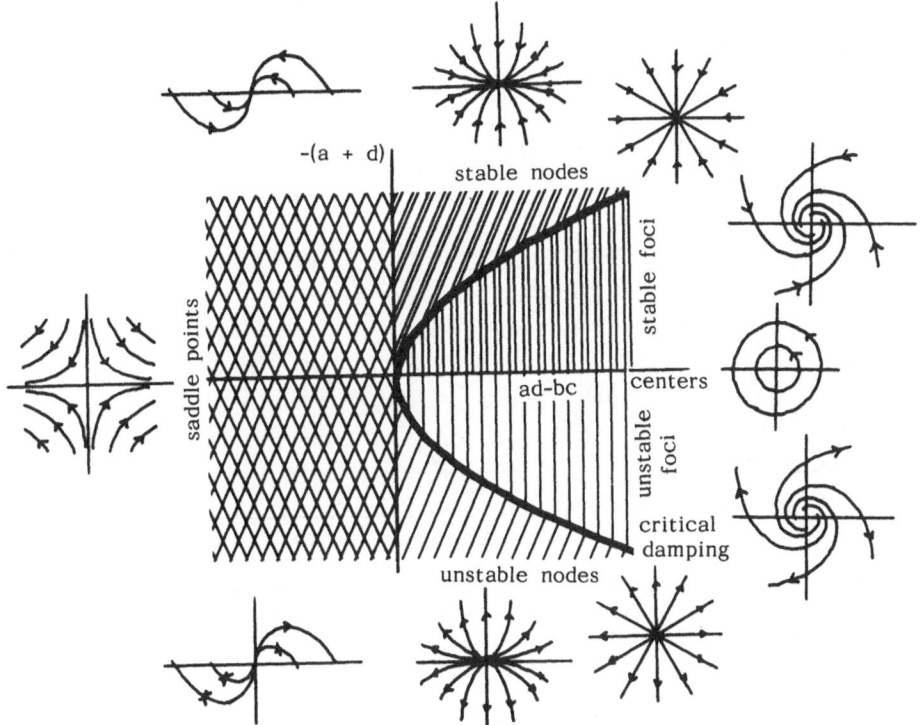

Fig.2.12: Relationship between type of equilibrium and coefficients of a pair of first order differential equations (equation 2.21). The type of equilibrium depends on the eigenvalues of the coefficient matrix of equation 2.21. The eigenvalues are given by the root

$$\lambda_1, \lambda_2 = \{ (a+d) \pm \sqrt{((a+d)^2 - 4(ad-cb))} \}/2$$

(e.g. HOCHSTADT, 1964; JACOBS, 1974; HADELER, 1974).

which provides allometric relationships for a wide range of parameter values (cf. Fig. 2.12).

Huxley's allometric relationship, therefore, appears as a rather likely first order approximation of the relationship between growing organs or measurements taken on a growing organism. However, there are numerous exceptions especially in ontogeny. Such an example is given in Fig. 2.13 -- the non-linear ontogenetic trend in a Paleozoic ammonite which, however, can be approximated by allometric relationships in different intervals.

2.2.2 The 'Ontogenetic Morphospace'

If one picks individuals of a certain age class from a species, then the morphological

24

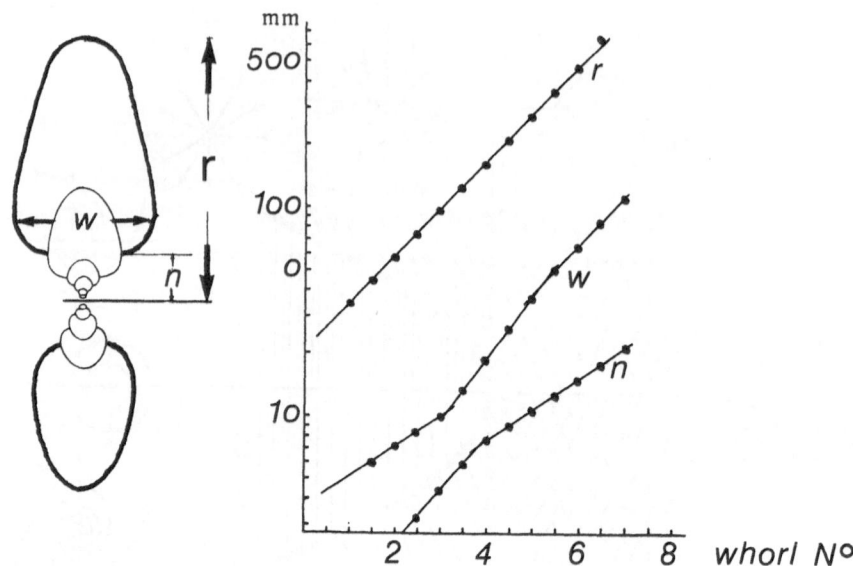

<u>Fig. 2.13:</u> Nonlinear ontogenetic relationships in a Paleozoic ammonite which can be stepwise approximated by simple allometric relationships (modified from KANT & KULLMANN, 1980).

features show usually a typical intraspecific variability, and in most cases the different features are correlated within every age class, e.g. size and weight are correlated and can be described by a two-dimensional Gaussian distribution for every age class. In the most simple case one needs two sources of variation to describe the ontogenetic mopho-space of a species:

a) for every age class a description of the variability of all features under consider-ation and their covariances. As a first approximation one can assume that the time sections through the ontogenetic morphospace are multi-dimensional Gaussian distributions;

b) a description how the mean of these distributions moves with increasing age through the morphospace. This gives a characteristic (mean) ontogenetic trace for the entire species -- for measurements, the mean (multidimensional) allometric relationship.

Fig. 2.14 illustrates this description of the morphospace whereby the 'mean ontogenet-ic trace' is approximated by a straight line (e.g. an ideal allometric relationship in loga-rithmic coordinates), and the age sections are idealized as ellipsoids (ideal Gaussian distri-bution). It is obvious that this description cannot be used only for continuous ontogenetic development (as in the ammonite example), it also holds for growth in finite steps like in crustacea. Thus, this kinematic model provides a relatively general description of the

25

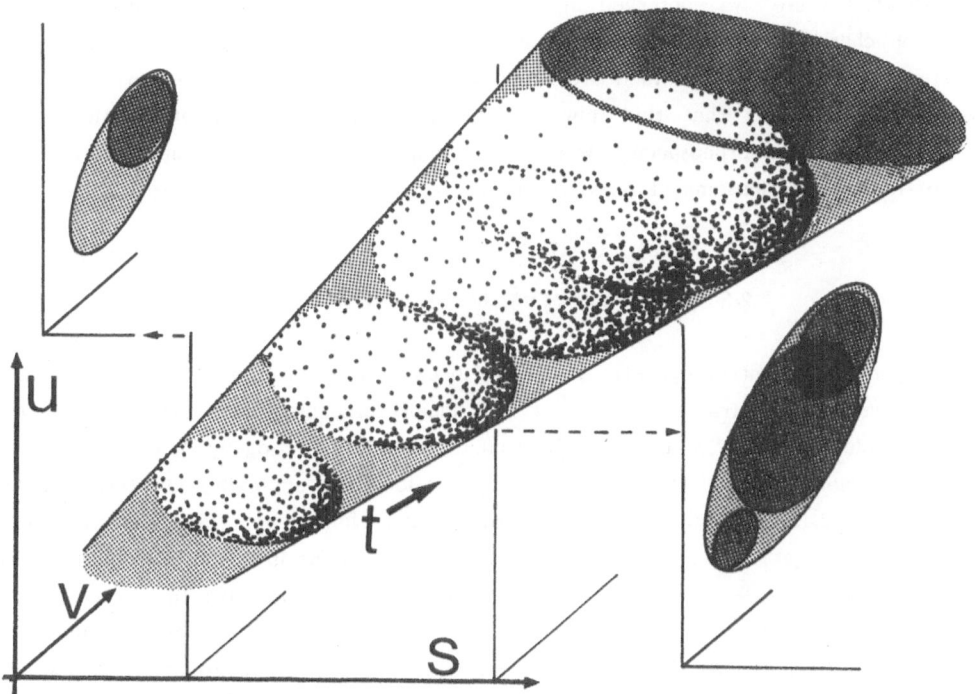

Fig. 2.14: A linear model for the 'ontogenetic morphospace' of a species. The variables u,v,s are parameters or measurements which characterize the morphology. The ellipsoids are time sections, i.e. they are the probability distributions for a certain age class. They are dislocated within the (u,v,s)-space with time t either continuously or in discrete steps. The hull of these ellipsoids (in the linear model a cone) is the probabilistic boundary of the ontogenetic morphospace. Sections through this morphospace by another variable than time, e.g. size (s), are ellipses which contain various age classes which may appear to be strongly correlated.

ontogenetic development as well as a definition of a probabilistic morphospace for the whole ontogeny.

If this ontogenetic morphospace is now sectioned by another variable than by age, e.g. by constant size which is an accessible control-parameter in paleontology -- then the section contains parts of the ontogenetic trend. Thus, even if the features under consideration are uncorrelated within an age-section, it is possible to find a strong correlation within the size--sections (Fig. 2.14). How strong this correlation will be, depends on the specific ontogenetic trace, on the correlation of features in the age-sections and on the angle between the principal axis of the age distribution and the ontogenetic trace.

'Buckman's law' was observed in those ammonites which show specially strong morphological changes through ontogeny, and the observed variability for a constant size consists   of morphotypes which are found as ontogenetic growth states in all specimens. Therefore, it is likely that this 'law' results simply from the oblique sections through the age dependent morphospaces; whereby a high correlation between features on the level of the age sections may inforce the strong correlation within the size sections.

2.2.3 Discontinuities in the Observed Morphospace

So far, the mean ontogenetic trace has been assumed to be a straight line or can be transformed into a straight line (i.e. if it is ideally allometric). However, even allometry is only an idealized first order approximation. Especially in ontogeny more complex relationships commonly occur,, which only allow an allometric approximation through certain intervals (Fig. 2.13, cf. KANT & KULLMANN; 1980). Non-linear relationships are usually found if morphology is described by some index numbers -- as it is the case in 'theoretical morphology' (e.g. RAUP, 1966). Thus, in the general case one has to expect that the mean ontogenetic trace is a three- or more-dimensional curve. This causes complications if time is not available as the controlling variable; e.g. the size sections will show deformations as a function of age. A mathematical description of the morpho-space without time can, therefore, lead to rather complicated nonlinear equations.

In addition, one has to expect complications in any projection of the n-dimensional ontogenetic morphospace (all possible relationships) onto a subspace, say the two-dimensional subspace of a point plot. Fig. 2.15 gives a sketch of such a curved ontogenetic morphospace. In the convex area of its hull a singularity appears due to the projection into

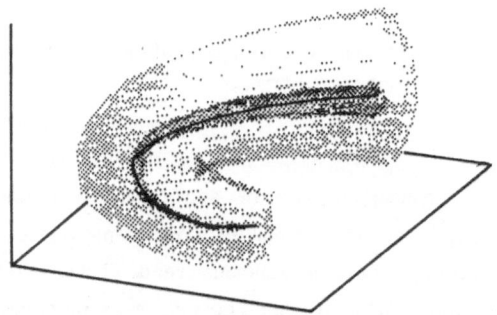

<u>Fig. 2.15:</u> The hull of a curved ontogenetic morphospace, a single ontogenetic trace and the probabilistic neighborhood of this trace. Other trajectories, which start close to the sketched trace, will be within this probabilistic neighborhood. In the concave area of the hull a swallowtail singularity appears, which will be discussed in chapter 4.

the plane. Such structurally stable singularities will be discussed in detail in chapter 4, however, some aspects of the deformations in subspaces can be already discussed here by the analysis of the ontogenetic traces of single specimens.

If one picks a certain set of ontogenetic traces for single specimens from the probabilistic ontogenetic morphospace, then, by experience, one can expect that they evolve in a regular manner and that they do not depart too much from their original relative position within the age section:

Experience shows that a juvenile 'pyknic' human will, in general, not turn into a 'leptosome' one during its ontogeny.

Now, we can describe the evolution of the ontogenetic morphospace as an iterated (or continuous) map which describes the change of the age dependent probability distribution and the dislocation of its mean. And, one can assume that the map, which generates the probabilistic ontogenetic morphospace of a species from some initial distribution, also describes the ontogenetic traces of single specimens up to some error term. If one neglects the error term, which causes the representation of ontogenetic traces by tubes rather than by lines (Fig. 2.15), then a significant regular disturbance within a family of ontogenetic traces can result only from the projection of the multi-dimensional space onto a subspace. What then reasonably can be expected, without further analysis, are local folds of the map (Fig. 2.2).

A simple model of such a fold in two dimensions is the tangent space of a parabola (Fig. 2.16c) whereby the tangents are local linear approximations of the ontogenetic trajectories. The concave side of the fold line, the parabola, is empty, no tangents pass through this area. In contrary, on the convex side of the fold line two tangents pass through every point of the plane. Naturally, such a fold model can be valid only as a local model. In this sense Fig. 2.16 provides a paleontological example of a local fold in the ontogenetic morphospace.

The ontogenetic traces of several individuals of the ammonite genus *Hyperlioceras* are drawn in a two-dimensional parameter space (non-allometric) which includes size (=Dm). The specimens belong to different species of this genus (BAYER, 1970), but this should not be a serious problem because the idea is only to show that local folds can be expected in paleontological 'growth' data -- under the aspects of the previous discussion these species may well be lumped into a single species. During late ontogeny, measured by size, an inversion of the morphological trend occurs (Fig. 2.16). Specimens with rather high relative whorl height (N) turn into forms with moderate values of this parameter and vice versa. This pattern is very regular with respect to the precision of the measurements, and the inversion occurs within a relative small size interval. Thus, the local behav-

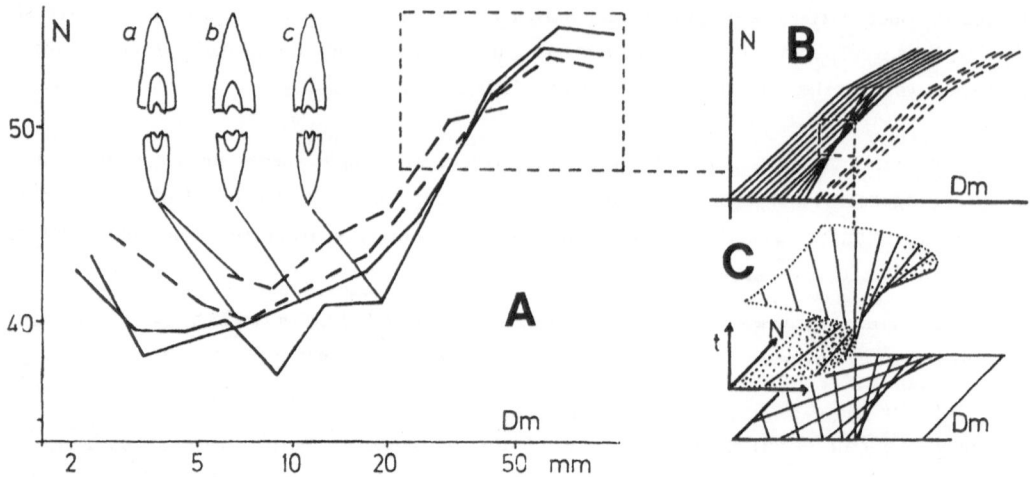

Fig. 2.16: Ontogenetic trajectories of ammonites of the genus *Hyperlioceras (a: H. desori, b: H. subsectum, c: H. discites)*, modified from BAYER (1969). N: relative height of whorl, Dm: diameter of the shell. During the late ontogeny, measured by size, an image inversion occurs, which can be interpreted as a local fold. In the model the fold causes local intersections of the trajectories and an empty area. If age (t) is used as an additional variable, one can expect that the trajectories are well separated, i.e. that the intersections are due to the projection onto the two-dimensional parameter space.

ior of the morphological trajectories can well be compared with a local fold. If age could be added as an independent variable, the trajectories would be lifted into the third dimension. However, if age is related to the earlier development in the parameter space (Dm,N), then the trajectories will be arranged in a more or less regular manner within the three-dimensional space (Dm,N,t). The local singularity, where the trajectories intersect, may then appear like a piece of a ruled hyperbolic surface (Fig. 2.16). The rulings model locally the ontogenetic traces, and their projection onto the (Dm,N)-plane is the discussed tangent space of a parabola.

This is not the place to say that this is the way to study and to describe the pattern of Fig. 2.16. But it is a way to illustrate and perhaps to overcome the difficulties which arise from singular structures like the regular intersection of the trajectories. In chapter 4 it will be shown that singularity theory or, more specific, elementary catastrophe theory provides a very elegant method to analyze such patterns. Anyway, it became clear that the ontogenetic development of morphology cannot always be considered to be linear, neither on the probabilistic level of the ontogenetic morphospace nor on the level of individual ontogenetic traces. The celebrated analysis of morphology by the so-called higher statistical methods (like factor analysis) has, therefore, to be used with caution. Patterns like in Fig. 2.16 cannot be linearized within the observed parameter space, and, therefore, they cannot be analyzed with linear models. On the other hand, the earlier discussion of the ontogenetic morphospace shows that, even within the most simple linear model, the ontogenetic trend cannot be ruled out for a linear factor analysis as is some-times assumed (BLACKITH & REYMENT, 1971). If the covariance structure is altered by age effects within the size sections, then we cannot reconstruct the original distribution from this sections without additional information -- in paleontology qualitative information will then be preferable against any quantitative measurement.On the other hand, the discussed models provide tools for the taxonomist. They give qualitative arguments for the variabili-ty of species and, therefore, for the definition of a species. In addition, they allow to formulate specific quantitative models.

2.3 ANALYSIS OF DIRECTIONAL DATA

The analysis of three-dimensional directional data by means of the 'stereographic projection' (Fig. 2.17) is a standard procedure in tectonics and sedimentology. The aim of the procedure is usually to estimate a density function of unknown form from data points on the sphere (cf. MARSAL, 1970). The reconstruction of the density distribution requires a smoothing process, in general a moving average. The classical hand method works with a counting area (circle) of 1% of the surface of the half sphere (or of its

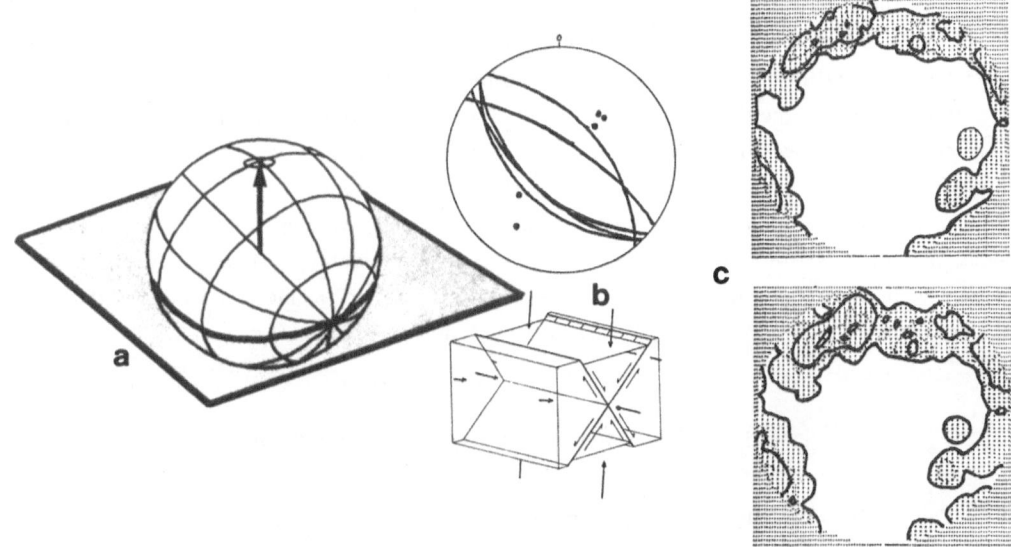

Fig. 2.17: a) Representation of a tangent plane in the unit sphere: by the 'circle of intersection' with the sphere, its unit normal and a point on the sphere (intersection of the normal with the sphere). b) a pair of idealized shear planes and a system of real shear plains in the stereographic projec-tion: representation by the 'circles of intersection' and the normals. c) Two stereographic projections of the same set of joints; above: Schmidt's grid (equal area); below: Wulf's grid (equal angles).

projection into the plane). When the first computer programs for the analysis of directional data appeared (e.g. SPENCER & CLABAUGH, 1967; ADLER et al., 1968; BONYUM
& STEPHENS, 1971; ADLER, 1970), they did not only simplify the analysis of directional
data, but they added new 'degrees of freedom': to choose the size of the counting circle,
to use various weighting functions or projections of the sphere (Fig. 2.17), and the computer allows to handle a rather large number of data. A question, which arose early
(KRAUSE, 1970), was, therefore, whether there exists an optimal size of the counting
area with respect to the number and to the distribution of data points on the sphere.
Alternatively, new 'influence functions' like an exponential decay function have been
introduced (BONYUM & STEVENS, 1971).

The problems associated with the smoothing process can be divided into more quantitative and more qualitative ones. The variation of the influence area (either by changing the diameter of the 'counting circle' or by different 'weighting functions') alters
the total number of expected values at a grid point. The classical way to standardize
this number to a percentage of all observed data points causes deformations of the
distribution in the way that the maxima are stretched -- the sum over all grid points
is greater than 100%. The counted data need to be normalized into 'densities per unit
area', or the area of influence has to be replaced by a weighting function for which
the integral over the area of influence equals one (BAYER, 1982). A more qualitative
aspect is that the smoothing process affects the variance of the distribution (GEBELEIN,
1951). This defect is mainly a function of the size of the area of influence. These problems are briefly discussed in the first section. However, while they are important in
a statistical sense, they are less significant for the geological interpretation of orientation data. In geology only the position of extrema may play a role for the structural
interpretation, and in this case the described deformations of the global distribution
do not affect the local interpretation. Therefore, most of the following discussion will
focus on the question whether the local extrema are stably estimated by the methods
currently in use. In the final section we will return to a more general problem and
analyze under which conditions we can suspect a density distribution at all.

2.3.1   The Smoothing Error in Two Dimensions

The estimation of a density function from scattered data on the sphere or its
projection onto the plane involves a moving average. For one-dimensional histograms
the resulting errors and the deformation of the moments have been discussed in detail
by GEBELEIN (1951). Fig. 2.18 illustrates how a one-dimensional histogram is deformed
if a moving average is used. Two-dimensional data and data on the sphere behave in
the same way (Fig. 2.19), and what we will do here is to estimate the error of the
smoothing process, i.e. the expected difference between the true and the computed
density distribution. Technically this requires Taylor expansions and integrations, however,

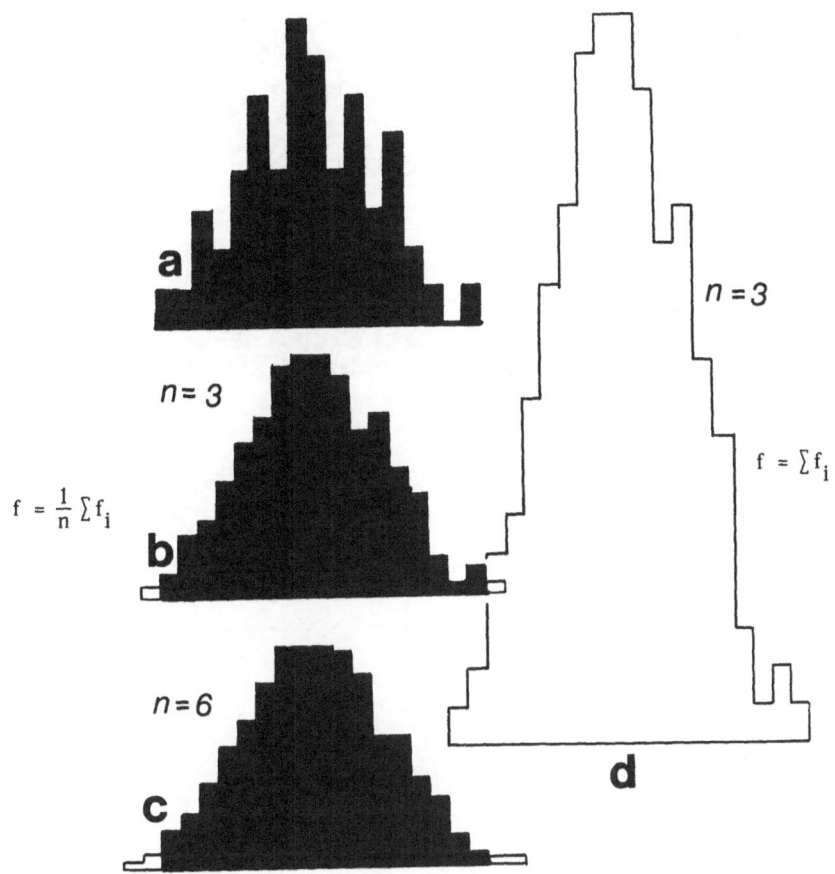

a

n = 3

n= 3

f = $\frac{1}{n}$ Σf_i

b

n = 6

c

n = 3

f = Σf_i

d

<u>Fig. 2.18:</u> Smoothing a histogram by a moving average: a to c: normalized averages; d: not normalized histogram of a three point moving average.

the mathematic remains rather simple.

The way to estimate the error is to compare the observed densities with a theoretical density function f(x,y) which is analytic (i.e. continuous and differentiable) with the values which result from averaging over a small interval. The error is the difference between the true value of the density function and the average. In the plane we choose an interval $(\Delta x, \Delta y)$ in the way that its center -- the arithmetic mean -- has coordinates (0,0). We can do this for any interval by simply shifting the coordinate system. To find the mean density within the interval we have to sum over all points within the interval and to divide by the area of the interval, i.e.

$$\bar{f}(0,0) = \frac{1}{\Delta x \Delta y} \, _Y\!\int _X\!\int f(x,y)dxdy; \qquad \begin{array}{c} \frac{-\Delta y}{2} \leq Y \leq \frac{\Delta y}{2} \\[2mm] \frac{-\Delta x}{2} \leq X \leq \frac{\Delta x}{2} \end{array} \qquad (2.23)$$

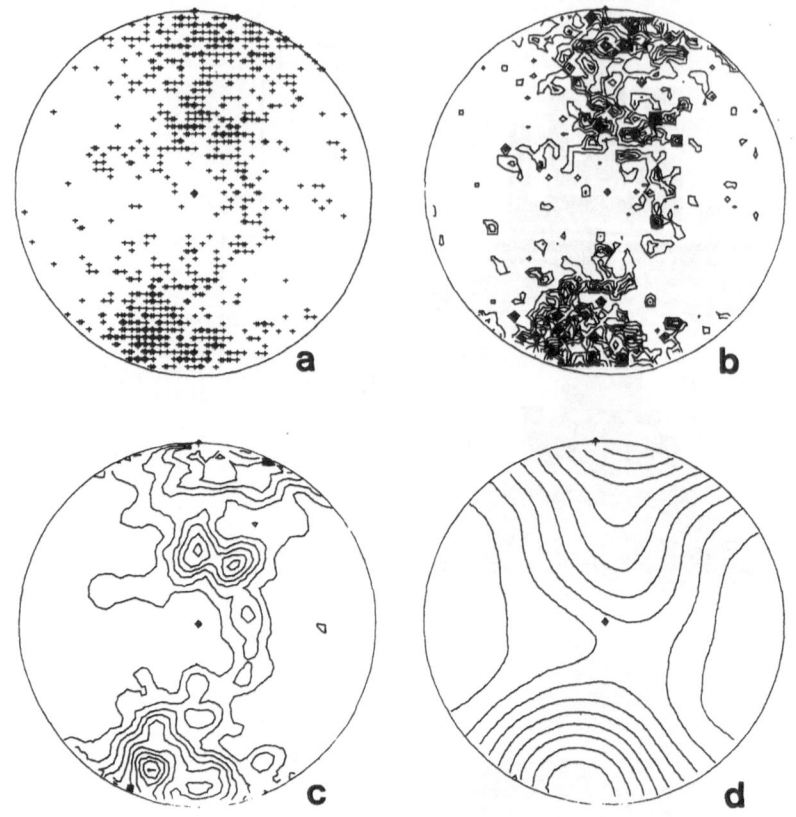

Fig. 2.19: Smoothing a 'histogram' (a) on the sphere. b-d: increasing 'area of influence' (triangular weighting function -- see text for discussion).

We assume further that the density function f(x,y) can locally be developed in a Taylor series:

$$f(0+h,0+k) = f(0,0) + hf_x + kf_y + (h^2 f_{xx} + 2hkf_{xy} + k^2 f_{yy})/2 + \cdots \qquad (2.24)$$

If f(x,y) in equation (2.23) is replaced by the approximation (2.24), we can evaluate the integral and find

$$\bar{\bar{f}}(0,0) = f(0,0) + \frac{\Delta x^2}{24} f_{xx} + \frac{\Delta y^2}{24} f_{yy} + \cdots \qquad (2.25)$$

whereby it is worth noting that all integrals over odd powers in equation (2.24) vanish, the largest error term, therefore, depends on the local curvature of the density function f(x,y) and is approximately

$$\delta f = f(0,0) - \bar{f}(0,0) \simeq -\left(\frac{\Delta x^2}{24} f_{xx} + \frac{\Delta y^2}{24} f_{yy} \right).$$ (2.26)

The error terms have a negative sign, i.e. maxima are lowered, minima are filled -- depending on the local curvature and on the area of influence. Thus, if we have enough information about the curvature of the density function, a good strategy would be to use a small element for averaging where the curvature is strong, and a large one where the function is flat, in order to combine a good approximation with fast computation. In the empirical problem, however, this will not be possible.

If the normalization by the area of the interval ($\Delta x, \Delta y$) is ignored, then it makes not even sense to speak about an error (cf. Fig. 2.17d), the averaging process then generates a totally new distribution as defined by equation (2.25). A safe strategy, which easily can be used by the computer, is to transform the initial data first into percentages (densities) and to normalize them again by the counting area after averaging.

In the case of a circular counting area the error is of similar form. In this case one has to evaluate the integral

$$\bar{f}(0,0) = \frac{1}{\pi \frac{x^2}{4}} \int_X \int_G f(x,y) \, dy \, dx; \qquad -\sqrt{(\Delta x^2 - x^2)} \le G \le \sqrt{(\Delta x^2 - x^2)}$$ (2.27)

or

$$\bar{f}(0,0) = \frac{1}{\pi r^2} \int_o^{2\pi} \int_{-r}^{r} f(r\cos\psi, r\sin\psi) \, r \, dr \, d\psi;$$

using again a Taylor approximation, the integral can be evaluated, and the error is of order

$$\delta f = f(0,0) - \bar{f}(0,0) \simeq \frac{d^2}{32} (f_{xx} + f_{yy})$$

where d: the diameter of the 'counting area'. (2.28)

Even for averaging on the sphere the error is of the same magnitude. Averaging on the sphere is easily evaluated with a computer if the data points are treated as vectors. The angular distance (a spherical cap) can be defined by the scalar product of vector pairs. The computation of the smoothing error proceeds as before, however, it is slightly more tricky to evaluate the integrals. We assume that the density function can be written as $f(x,y) = g(x,y,z)$; because we are dealing with unit vectors, we have $z = (x^2 + y^2)^{1/2}$. Thus, we can again use the Taylor expansion (2.24), and the mean density within the area F (area element df) can be written

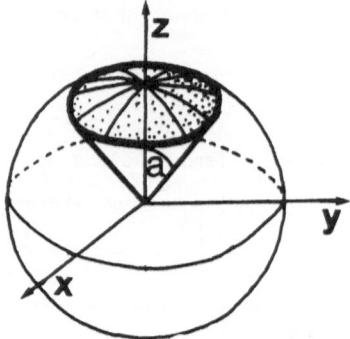

Fig. 2.20: The 'area of influence' on the unit sphere after its center has been rotated to coincide with the z-axis (see text for discussion).

$$\bar{f}(0,0) = \frac{1}{F} \{ f(0,0)\int^F df + f_x \int^F hdf + f_y \int^F kdf $$

$$+ \frac{1}{2}(f_{xx}\int^F h^2 df + 2f_{xy}\int^F hkdf + f_{yy}\int^F k^2 df) + \ldots \} \quad . \qquad (2.29)$$

Transforming the coordinate system to spherical coordinates yields the integral

$$\bar{f}(0,0) = \frac{1}{2\pi(1-\cos a)} \{ f(0,0)\int_0^{2\pi}\int_0^a \sin\delta d\delta d\psi +$$

$$+ f_x \int\int \cos\psi\sin^2\delta d\delta d\psi + f_y \int\int \sin\psi\sin^2\delta d\delta d\psi$$

$$+ \frac{1}{2}(f_{xx}\int\int \cos^2\psi\sin^3\delta d\delta d\psi \qquad (2.30)$$

$$+ 2f_{xy}\int\int \cos\psi\sin\psi\sin^3\delta d\delta d\psi$$

$$+ f_{yy}\int\int \sin^2\psi\sin^3\delta d\delta d\psi) + \ldots \} \quad .$$

The coordinate system has been chosen so that the centroid of the spherical 'counting cap' coincides with the z-axis of the sphere. The angle 'a' defines a cone with the z-axis its central line (Fig. 2.20), whereby all integrals containing cos a and sin a vanish. The intersection of this cone with the sphere bounds the averaging area. The estimated density value is of magnitude

$$\bar{f}(0,0) = \frac{1}{2\pi(1-\cos a)} \quad f(0,0)2\pi(1-\cos a)$$

$$+ f_{xx}\pi(\frac{2}{3} - \cos a + \frac{1}{3}\cos^3 a) \qquad (2.31)$$

$$+ f_{yy}\pi(\frac{2}{3} - \cos a + \frac{1}{3}\cos^3 a)$$

or $$\bar{f}(0,0) = f(0,0) + (f_{xx}+f_{yy})(\frac{1}{3} - \frac{1}{6}\cos a(1 + \cos a)).$$

The averaging error is approximately

$$\delta\bar{f}(0,0) = (\tfrac{1}{3} - \tfrac{1}{6}\cos a\,(1 + \cos a))(f_{xx}+f_{yy}).$$ (2.32)

2.3.2 Stability of Local Extrema

The major problem of a geological analysis of directional data is not so much the "statistical stability", but it is the stability of local extrema. However, in this context it turns out that the classical approach causes problems, i.e. the smoothing process by a step function over an influence area or by a rectangular weighting function. Let the area of the rectangular weighting function increase until it reaches the surface area of the half sphere, then the whole distribution becomes equalized, independent of the original data pattern. In other words, the extrema are smeared out with increasing area of influence. The instability of the local maxima becomes more obvious if one studies sparse data structures. If the area of influence, the counting circle, is varied on such a data set (Fig. 2.21), then new maxima are generated whenever two areas of influence

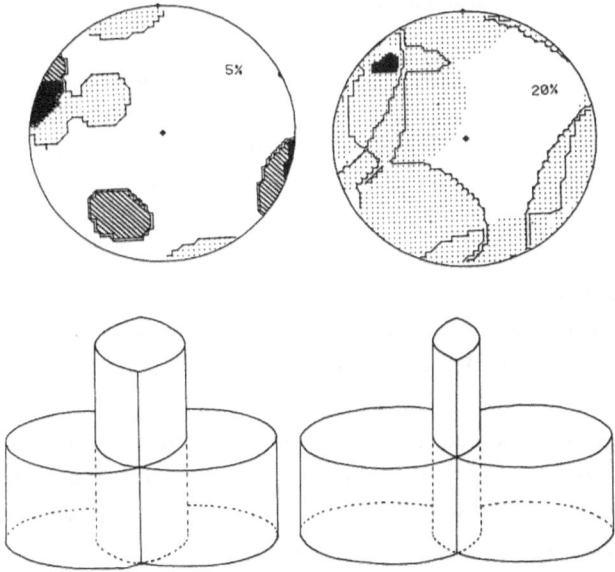

Fig. 2.21: Two stereographic projections for the same sparse data pattern but with different sizes of the 'counting circle' or 'rectangular weighting function' (5% and 20% of the area of the half sphere). The location of maxima (dark areas) depends strongly on the size of the counting circle. The lower figures illustrate how the maxima arise over the intersection area of the rectangular weighting functions ('counting circles').

begin to overlap. Furthermore, the transition to such a new maximum is not a smooth process, but it is a 'sudden jump'. The smoothing process is, therefore, highly instable with respect to the position of the local maxima and to small size changes of the in-fluence area.

However, it is not only the size of the influence area which may cause the sudden occurrence and disappearance of local maxima. The smoothing process is executed on a finite net either on the sphere or on a two-dimensional projection of the sphere. There-fore, any change of the grid structure will locally change the overlapping pattern of the influence areas, and the identical instabilities will arise. Because there exists no 'equally spaced grid' on the sphere, even a rigid rotation of the data may alter the distribution of maxima. Thus, the position of local maxima is not invariant against (even small) changes of the grid structure or of the size of the counting circle. Indeed, the classical method has only two stable states with respect to local extrema:

a)   the situation where the counting area is identical with the surface area of the half sphere, in which case all extrema vanish, and therefore all information is lost,

b)   the case where the counting area goes to zero and the distribution resem-bles a plot of the original data points. This case, however, does not summa-rize the structural information.

Now, one could go back to the old hand method and hope that all problems can be solved by a 'standardized' counting area and grid structure because then the system is forced to be without variations or irregularities. However, this would only cover the problem because any addition or subtraction of a data point (or of a group of data points) can change the local pattern of extrema in the same way as discussed above. The reason is that the method does not preserve a data point as the smallest possible maximum, but that the maxima appear in the intersection areas of the counting circles (Fig. 2.21). The equivalence between the grid and the data can be illustrated by two strategies, which can be used for computer-algorithms (Fig. 2.22):

1) The classical method treats a data point as a point and defines a grid on the sphere (or its projection). To every grid point a counting circle is assigned, and one has to find the number of data which fall within this circular area, i.e. all points are summed up at a grid point which are within a distance limit.

2) Another strategy is to assign the counting circle to every data point. Then one has to find all grid points which fall inside the distance limit (which surrounds the data point) and to add the   (weighted)   value of the data point to the grid points which satisfy the distance property.

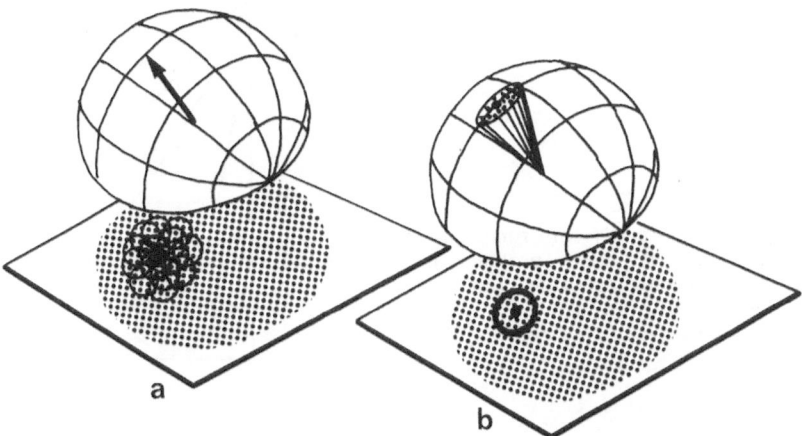

Fig. 2.22: Two viewpoints of the smoothing process. a) A data point belongs to a grid point with a certain probability. Several grid points are in 'competition'. b) One or several data points have a certain probability to be correctly located -- the spherical cap, which is associated with a data point, is an area of confidence. In the projection the grid points have to be found which belong to the area of confidence.

Both strategies are possible implementations. In the first case, one usually has to store all data in the central memory while the grid points can be treated one after the other; in the second case, one stores the grid and can call one data point after the other from some external memory. It is this second method which illustrates that the data points are not preserved as the smallest possible maximum. Actually, they are replaced by a rectangular (density) function over a finite area. Variation of the area of the counting circle, thus, resembles the disturbance of a signal (Fig. 2.1). The counting circle defines the area in which the data point will be found with a certain probability, and this probability is everywhere equal (within the area of influence). The rectangular weighting function over the counting circle, therefore, can be interpreted as a rectangular probability distribution. A consequence is that one can not either assume that another (sparse) sample from the identical universe shows the same extrema.

It turnes out that the smoothing process by a rectangular weighting function is very sensitive and locally highly unstable with respect to the position of the extrema. The striking point is that other weighting functions give much better results with regard to the considered stability problem. Such a family of non-standardized weighting functions, which includes asymptotically the point plot and the rectangular weighting function, is given by the family of polynomials (Fig. 2.23)

$$w(r) = (1 - r)^n; \qquad 0 \leq r \leq 1$$

$$w(r) = 0 \qquad\qquad r > 1$$

$$(2.33)$$

$$r = R/R_0, \quad n = 1,2,3,\ldots, 1/2, 1/3,\ldots$$

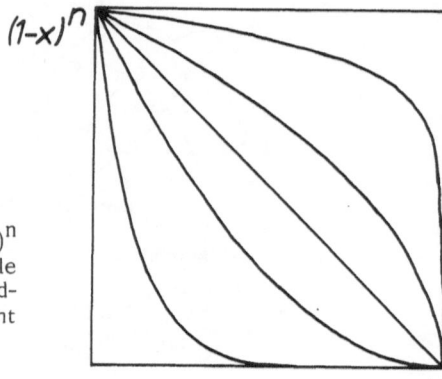

$(1-x)^n$

Fig. 2.23: The family of functions $y=(1-x)^n$ in the interval $0 \le x \le 1$ provides a possible family of weighting functions (not standardized). They include asymptotically the point plot and the rectangular weighting function.

where R_o is the diameter of the counting circle, and R is the distance of a grid point from the data point. For every grid point one has to form the sum Σ w(r) over the data points which are inside the distance R_o. In the case of a rectangular weighting function one has w=1 within the counting circle and w=0 outside. For the point plot one has $r_o=0$, w=1 at the data point. To normalize the estimated distribution one has to sum over all counts and then to divide the grid point values by this total sum.

Figs. 2.24 and 2.25 illustrate that the major maxima are well preserved, even for large counting areas if one of these functions is used, especially if the exponent 'n' is chosen small enough -- while for large 'n' the weighting function resembles the rectangular weighting function. A similar result was found by BONYUM & STEVENS (1970). They used an exponential decay function as a weighting function which was defined over the entire grid area, i.e. every data point contributed to every grid point. Their computer studies showed that the resulting frequency distributions are as useful as those generated by the classical 'hand' method. The point is that the stability of the local maxima increases if the weighting function has a well pronounced maximum at the center of the counting circle in local coordinates or at the data points in global coordinates. As this maximum of the weighting function vanishes -- like in the case of a rectangular weighting function -- the estimated extrema become instable in the sense that they do not further resemble the original data points. Fig. 2.26 illustrated how a triangular weighting function preserves the local data structure in contrast to the rectangular weighting function of Fig. 2.21.

In summary, it seems worthwhile to discuss the two technical strategies given above in more general terms. If the counting circle is associated with the grid points, then we are working in local coordinates. The question is whether a data point belongs with some probability to this grid point. In general, one assumes that this probability

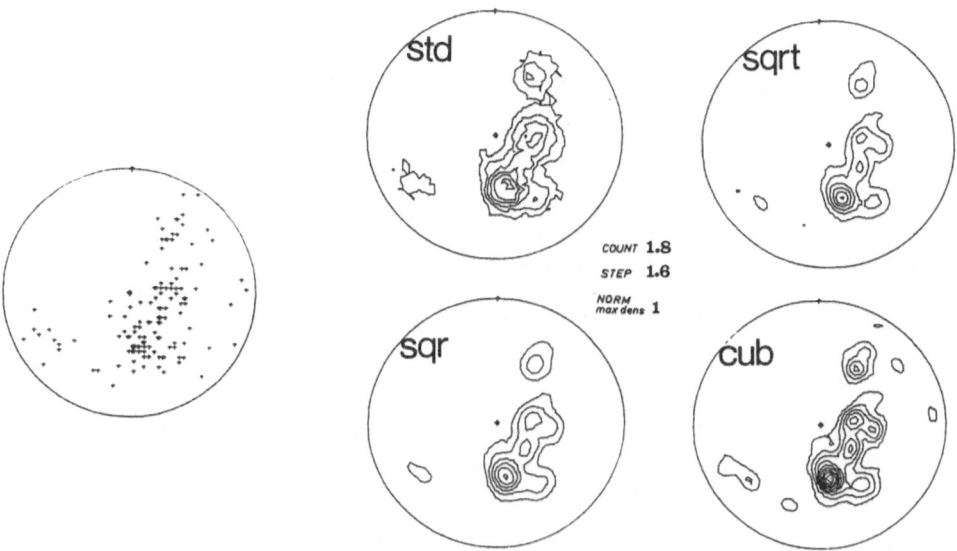

Fig. 2.24: Estimated frequency distributions on the sphere by use of various weighting functions: point plot (data); std: rectangular weighting function (classical counting circle);

sqrt: $w=(1-r/r_o)^{1/2}$;   sqr: $w=(1-r/r_o)^2$;   cub: $w=r-r^2(3-2r)$.
The size of the counting circle is 1.8% of the surface area of the half sphere for all smoothed distributions.

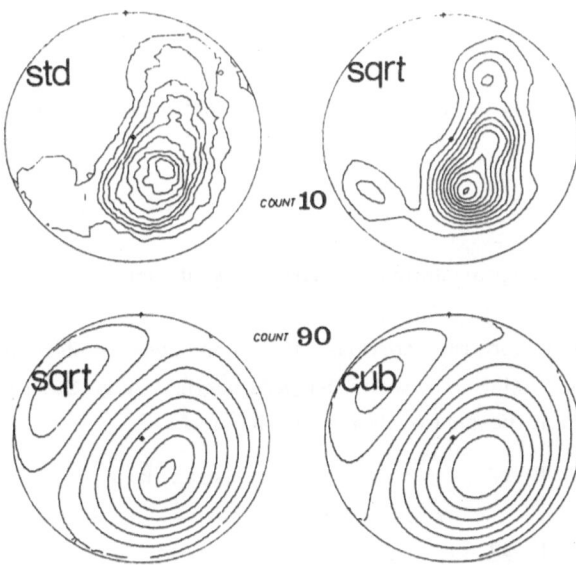

Fig. 2.25: Estimated frequency distribution on the sphere for counting circles of different size and for various weighting functions. Data and symbols like in Fig. 2.24. Counting areas 10% and 90% of the surface area of the half sphere.

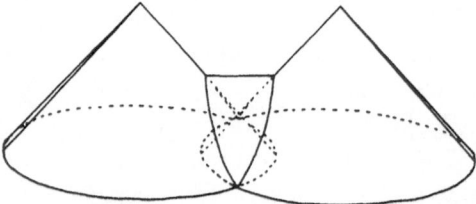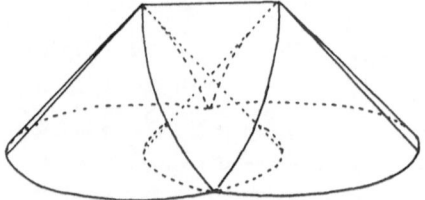

Fig. 2.26: Triangular weighting functions over a circular area of influence preserve the data point as a local maximum. As the overlapping of the circles of influence increases, either by increasing density of data points or by a larger area of influence, the new maxima develop smoothly between the grid points.

decreases continuously with the distance from the grid point. The other possibility is to assign the counting circle to the data point. In this case, we are working in global coordinates, i.e. we have to find the grid points which fall inside the counting circle. The question is now what is the probability for the data point to fall on a specific grid point -- and again the probability should decrease with increasing distance between the data point and the grid point. In any case, the smoothing procedure focuses on the "sampling error", i.e. the problem that we have only a small sample from a large universe of usually unknown structure.

In a statistical sense, the weighting function should assure that a random sample, which is taken from the smoothed distribution, will not depart too much from the original data set, at least if the estimated distribution is corrected for the variance (GEBELEIN, 1951). MARSAL. (1970) even tried to introduce a test-statistics on this argument. For the rectangular weighting function this condition does obviously not hold, as Fig. 2.21 illustrates..

2.3.3 Approximation and Averaging of Data

Another useful aspect for the previous discussion can be developed from the methods for local surface fitting. The local Shepard method (e.g. SCHUMAKER, 1976) uses a weighting function for the approximation of surfaces:

$$w(r;R) = \begin{cases} 1/r & ; \ 0 < r \leq R/3 \\[2mm] \dfrac{27}{4R}\left(\dfrac{r}{R} - 1\right)^2 & ; \ \dfrac{R}{3} \leq r \leq R \\[2mm] 0 & ; R < r \end{cases} \tag{2.34}$$

and the local estimation at any point (x,y) is given by

$$f(x,y) = \begin{cases} \dfrac{\sum F_i (w(r_i;R))^\mu}{\sum (w(r;R))^\mu} & ; \ r_i \neq 0 \\[4mm] F_i & ; \ r_i = 0 \end{cases} \tag{2.35}$$

where the F_i are the observed z-values at the data points (or frequencies); r_i is the distance between a grid point and a data point; μ : is a parameter (a 'metric') which can be freely chosen; R: is the radius of the area of influence (the 'counting circle').

Formula (2.35) is defined at all points (x,y) in the plane. It interpolates the observed values correctly at the data points (the values F_i at $r_i=0$) while the values at non-data points are weighted averages of the data points which lie within a distance R of the grid point. The weights are defined by equation (2.34). The local Shepard method provides an approximation method which is based on a 'counting circle' as discussed above. SCHUMAKER (1976) shows that the local Shepard method is an optimal approximation strategy for the local surface reconstruction. The relatively complicated definition of the weighting function ensures that the observed F_i-values, i.e. the data points, are preserved.

The local surface fitting method can be used to estimate a density function if the samples become rather large. For a large sample, with space coordinates measured on a discrete scale (with fixed precision), one can expect that the resulting histogram is a good estimation of the statistical universe. The local surface approximation (with normalization over the area of influence R) gives a first order approximation of the density function. As the sample size decreases, the histogram becomes noisy, and the local surface approximation produces local extrema which are due to the sampling error. A smoothing process is then required, and this procedure should be capable to eliminate the sampling noise. On the other hand, the smoothing process should not deviate too much from the local approximation. If one takes Shepard's method as a model for the local approximation, then the transition to a smoothing procedure requires to drop 1/r term for the interval $0 < r < R/3$ in equation 2.34 because this term would cause infinite values (if $r_i=0$). The remaining parts are

$$w(r) = \begin{cases} (\dfrac{r}{R} - 1) & ; \ 0 \leq r \leq R \\[4mm] 0 & ; \ R < r \end{cases} \tag{2.36}$$

This is the earlier discussed polynomial weighting function (2.33), which together with the first term in equation (2.35) provides a smoothing procedure. It is not hard to see that the rectangular weighting function is the most degenerated case of the family of weighting functions defined by equation (2.36); it is approached as $\mu \to \infty$.

Another question is how to choose the parameter μ, or which function of the family $w(r;\mu)$ is optimal. In order to see what we can achieve by the parameter μ of equation (2.36) or by the parameter 'n' of equation (2.33), one may discuss the most general form of these weighting functions:

$$w(x) = (1 - x) ; \quad x \leq 1$$
$$x = r/R .$$
(2.37)

Equation (2.37) takes the values $w(r=0)=1$ at the data point and $w(r=R)=0$ at the boundary of the area of influence, as required. The first derivative takes the values

$$w'(x) = -\mu(1 - x)^{\mu-1} \quad \text{with} \quad \mu \geq 0$$
$$\text{with } \mu \geq 0$$
$$w'(0) = -\mu$$
$$w'(1) = -1 \text{ if } \mu=0 \quad \text{else } w'(1) = 0 .$$
(2.38)

The parameter μ allows to adjust the slope of the weighting function at the grid point. However, to use this additional degree of freedom requires additional information about the proper value for the slope at x=0. From a general viewpoint, therefore, the straight line approximation (n=1) is nearly optimal (DeBOOR, 1978). It does not require any additional assumptions, and it solves the problem to connect the data point with the boundary of the area of influence by a continuous function.

However, a natural boundary condition for the weighting function could be that the first derivatives vanish at the data point and at the boundary of the counting circle. If one considers the weighting function as a probability distribution which assigns a data point to a grid point, then the Gaussian distribution would, of course, be a model with an infinite area of influence. In terms of polynomial weighting functions this condition cannot be satisfied by equation (2.37), at least we need a cubic polynomial like

$$w(x) = (x^3 + ax^2 + bx + c).$$
(2.39)

The boundary conditions $w(0) > 0$, $w(R) = 0$, and $w'(0)=w'(R)=0$ are only satisfied by the polynomial

$$w(x) = x^3 - \frac{3}{2}x^2 + \frac{1}{2}$$
(2.40)

or

$$w(x) = s(1 + x^2(2x - 3)).$$

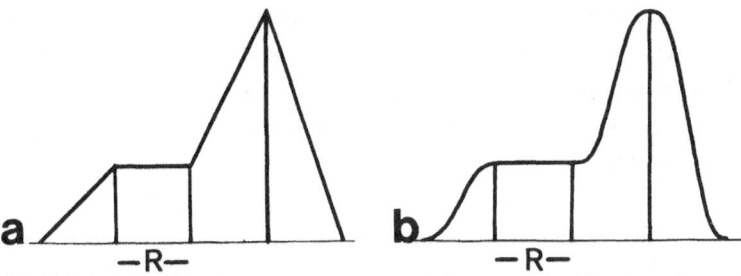

Fig. 2.27: Interpolation by spline functions: a) the linear Euler spline; b) the cubic spline with vanishing derivatives at the grid point and the boundary of the area of influence (R).

It turns out that there is a limited number of optimal weighting functions depending on the boundary conditions (Fig. 2.27). Equation (2.39) allows alternatively to adjust the slope of the weighting function in an arbitrary way at the grid point and at the boundary of the area of influence. It is the cubic Hermite interpolation and can be used for cubic spline interpolation (De BOOR, 1978). The earlier discussed triangular weighting function is simply the linear Euler spline.

The linear and the cubic spline provide a simple interpretation of the radius of influence (R). Assume the situation of two data points with distance R: Without loss of generality we can locate them on the real line and situate one of them at the origin, the other at the point x=1 (i.e. x=r/R). The weighting functions for the two points are

$$w_1(x) = (1 - x)$$
$$w_2(x) = (1 - (1 - x)) = x$$

(2.41)

for the linear spline and

$$w_1(x) = (1 + x^2(2x - 3);$$
$$w_2(x) = (1 + (1-x)^2(2(1-x) - 3) = x^2(3 - 2x)$$

(2.42)

for the cubic spline. In both cases we find that $w_1+w_2=1$ within the interval (0,R). The approximated values within this interval are

$$f(x) = f(0)w_1 + f(1)w_2$$
$$= f(0)(1 - w_2) + f(1) w_2$$
$$= f(0) + (f(1) - f(0))w_2 ,$$

(2.43)

that is a straight line in the case of the linear spline and a cubic function in the second case. However, if f(0)=f(R) (e.g. single measurements at both points), then the connection between the two data points is simply the horizontal line f(0), cf. Fig. 2.27. The radius of influence R, therefore, defines a threshold of resolution: Data points with distance less or equal R are subsumed in a single maximum while data points with distances larger than R are preserved as distinct maxima. As far as it is possible to relate R to the number of data (cf. KRAUSE, 1970), the spline functions provide optimal approximation and smoothing capabilities.

It is instructive to invert the above argument. If one defines the threshold property as the optimal solution, then one can discuss any polynomial weighting function in terms of the optimal solution, i.e. we have to find coefficients so that $w(x)+w(1-x) = $ constant. In the case of an arbitrary polynomial

$$w_1 = a_0 + a_1 x + a_2 x^2 + \ldots + a_n x^n$$
$$w_2 = a_0 + a_1(1-x) + \ldots + a_n(1-x)^n \tag{2.44}$$

the condition can only be satisfied if $a(-x)^n = -ax^n$, that is, if the leading power term is an odd number. The polynomials of equations (2.33) and (2.36), therefore, are not optimal. Now, we may analyze the special case of the cubic weighting function:

$$w_1 = ax^3 + bx^2 + cx + d$$
$$w_2 = a(1-x)^3 + b(1-x)^2 + c(1-x) + d$$
$$= -ax^3 + (3a+b)x^2 - (3a+2b+c) + \cdot(a+b+c+d). \tag{2.45}$$

To satisfy the condition all terms which contain powers of x have to vanish in the sum w_1+w_2. This gives the following relations between the coefficients:

$$
\begin{aligned}
b + (3a+b) &= 0 \quad \dashrightarrow \quad 3a = -2b \\
c - (3a+2b+c) &= 0 \quad \dashrightarrow \quad 3a = -2b,
\end{aligned}
\tag{2.46}
$$

and we have twice the same relationship between 'a' and 'b' which, of course, appeared already in equation (2.39). The parameter 'c' is arbitrary -- if $a=b=0$ and $c\neq0$, the cubic function degenerates to the linear Euler spline.

The general solution for a cubic spline under the condition $w(x,R)+w(R-x,R)=$constant is just the linear combination of the special cubic spline of equation (2.39) and the linear Euler spline. We can rewrite equation (2.45) with the parameters of equation (2.46) as

$$w(x) = \{a(x^3 - \frac{3}{2}x^2) + \frac{d}{2}\} + \{\frac{d}{2} \pm cx\}, \tag{2.47}$$

i.e. as a linear combination of the spline functions. To see what happens at the special points x=0 and x=R we take the derivative of equation (2.47) at these points

$$w' = 3ax^2 - 3ax \pm c$$
$$w'(0) = \pm c \qquad\qquad (2.48)$$
$$w'(1) = \pm c$$

and find that the weighting function has identical slopes at the points under considera- tion. Some weighting functions, which satisfy the discussed conditions, are given in Fig. 2.28, and it turns out that these functions are no proper weighting functions as they

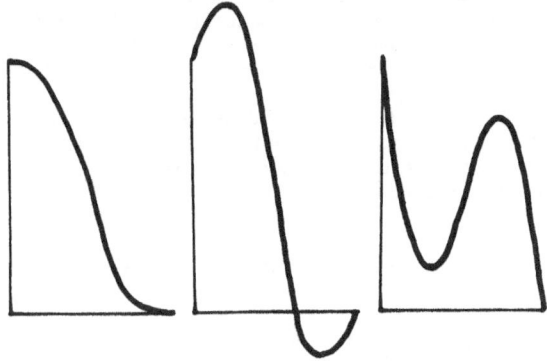

Fig. 2.28: Symmetric cubic weighting functions which satisfy the condition w(x)+w(1-x)=constant.

have extrema inside the 'counting circle' and may even assume negative values. The only remaining weighting function is the cubic spline of equation (2.39), however, this function tends to produce local 'platforms' at the data points (cf. Fig. 2.27). Thus, the linear Euler spline remains the optimal solution for a smoothing procedure.

2.3.4 A Topological Excursus

The previous discussion focussed on the problem to find a proper weighting function which takes values greater than zero inside a certain area and the value zero outside this area. A similar problem occurs in topology and is solved there by Urysohn's lemma -- for details see JÄNICH (1980). Formally we can formulate our problem in the following way (cf. Fig. 2.29):

What we are looking for is a function f: x→ $[0,1]$ on U ⊂ V ⊂ W which has value 1 on U and value 0 on W. On V = W\U we are looking for a continuous connection. The simple idea is (JÄNICH, 1980) to construct a step function which

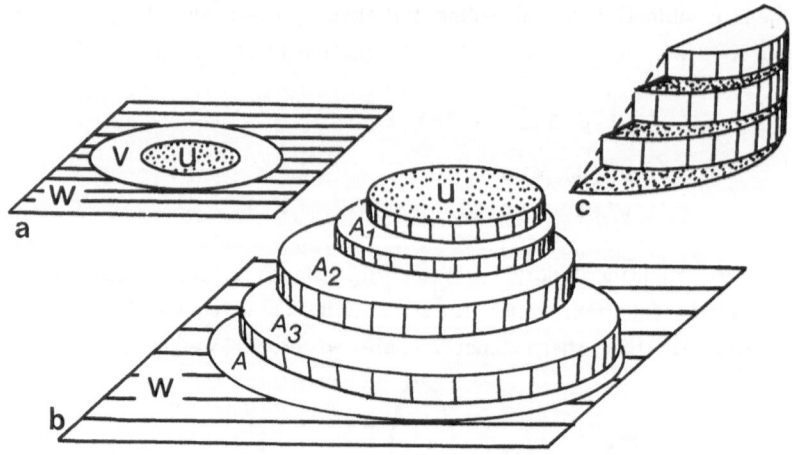

Fig. 2.29: (a) The problem is to find a continuous connection between the areas U and W. (b) A possible solution is to construct a step function in V = W U. (c) However, if the boundaries of the subsets A_1, A_2, ... are locally in contact, a further refinement of the step function is not possible.

decreases from U to W (Fig. 2.29 B), i.e. one has to find a chain of sets

$$A = A_1 \subset ... \subset A_n \subset W \backslash U.$$

The step function takes the values

1 on U,   1 - i/n on A_i and 0 on W.

By stepwise refinement one can construct the continuous connection between V and W. The only problem is that the boundaries of U_i and U_{i-1} do not meet (Fig. 2.29 C). In this case, it would not be possible to construct a continuous connection, i.e. we cannot insert an additional step between the boundaries (JÄNICH, 1980).

If one applies this to the discussion of weighting functions, it becomes immediately clear that the rectangular weighting function is the extreme solution which does not allow any further refinement between the areas V and W because U = V\W = $\emptyset$. On the other hand, the linear Euler spline is a possible connection, even when V shrinks to a point.

2.3.5 Densities, Folds and the Gauss Map

So far, the reconstruction of density distributions was discussed without regard to the question whether there exists a density distribution for orientation data. In the case of current oriented obstacles in sedimentology and of lineations and cleavages in tectonics, we can use standard arguments. One can expect that in these cases the orientation is controlled by a potential (BAYER, 1978). The systems stay at the minima of the potentials, i.e. in the position of minimal drag in a current; cleavage will occur in the direction of maximal shear stress (minimal normal stress) etc. To arrive at a probability distribution or a density distribution, one assumes that the objects are driven out of the minimum position by random forces. One way to find the density distribution is provided by the stationary solution of the Fokker-Planck equation (e.g. HAKEN, 1977).

The situation is different when orientation data are taken from surfaces, e.g. from deformed bedding planes in tectonics. The orientation data are then the normals of the surface, and therefore the surface needs to be regular, that is, there exists a tangent plane at every point of the surface. If the surface is given as a map $R^2 \to R^3$, e.g. as

$$X(u,v) = \{x(u,v), y(u,v), z(u,v)\},$$

(2.49)

then the unit normal vector at each point p of the surface is given by

$$N(p) = (X_u \wedge X_v)/(|X_u \wedge X_v|);$$

$\wedge$ is the vector product.

(2.50)

Now, in differential geometry the map $N: S \to R^3$ (S: the surface) is studied which takes its values on the unit sphere

$$S^2 = \{(x,y,z) \ R^3 \mid x^2+y^2+z^2 = 1\}.$$

(2.51)

The map $N:S \to S^2$ is called the Gauss map of the surface S (DoCARMO, 1976), and this map is equivalent to the standard representation of orientation data on the unit sphere. The Gauss map has various properties, which can be interesting for geological interpretation of directional data. Detailed discussions are given by DoCARMO (1976) and DANGELMAYR & ARMBRUSTER (1983).

The first point to be discussed is how a surface element 'S' maps onto the unit sphere. If only local properties of a surface are considered, that is a small neighborhood of a surface point, then typically three situations occur -- the local surface structure is either elliptic, parabolic or hyperbolic as illustrated in Fig. 2.30. With regard to the positive normals of the surface one finds that the Gauss map preserves orientation at an elliptic point and reverses it at a hyperbolic point (Fig. 2.30) -- at a parabolic point a degenerated situation arises, the normal vectors are all aligned in a plane which intersects the sphere. In geological problems one has also to consider the inverse surfaces (synclines). In this case, the closed pathways around the surface points in Fig. 2.30 are inverted in the Gauss map.

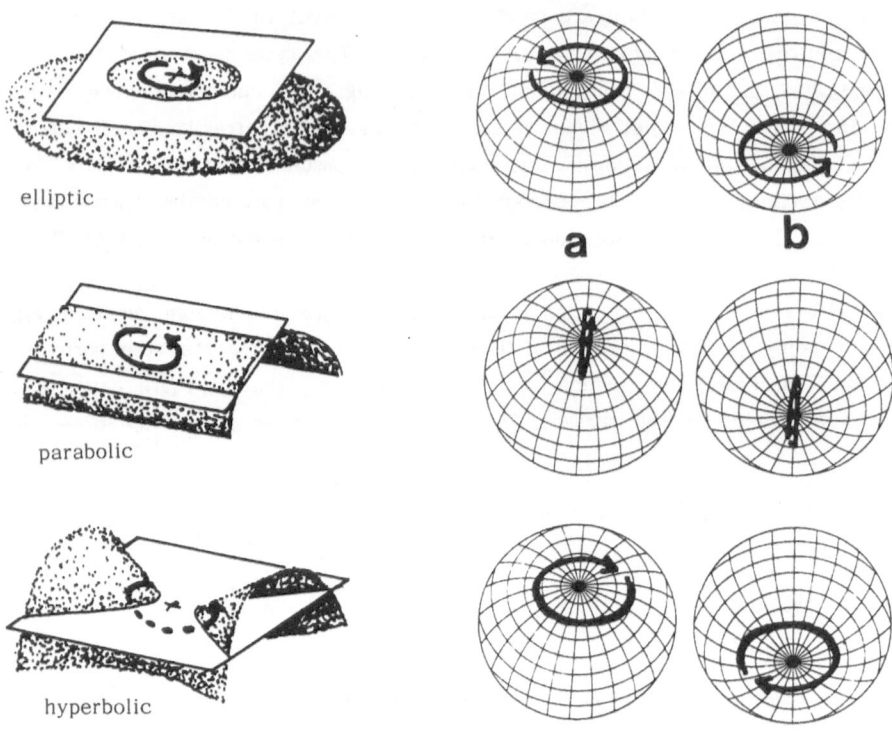

Fig. 2.30: The local structure of a surface is either of elliptic, parabolic or hyperbolic type. The orientation of a closed pathway (surrounding the critical point) is preserved at an elliptic point and reversed at hyperbolic and parabolic points (a: positive normals, anticlines). For synclines (b) the entire pattern is reversed. Alternatively, (a,b) can be interpreted as projections in the upper and lower hemisphere -- both methods are used in geology. Grids: inclined Lambert's equal area projection; adapted from HOSCHEK, 1969).

After these geometrical considerations we analyze how a surface element 'S' maps onto the unit sphere, and which value the area of the surface element takes on the unit sphere. If we choose a very small surface element dB, the expected density of normals on the sphere is proportional to dB/dS (where dS is the corresponding surface element on the sphere). The area of a small surface element is

$$B = \iint_R |X_u \wedge X_v| \, du \, dv, \tag{2.52}$$

and the image of B on the unit sphere has area

$$S = \iint_R |N_u \wedge N_v| \, du \, dv. \tag{2.53}$$

(R is the area in the (u,v)-plane which corresponds to the surface element B). 'S' can be expressed by

$$S = \iint_R K |X_u \wedge X_v| \, du \, dv \qquad (2.54)$$

where K is the Gaussian curvature of the surface. The relation between the surface elements B and their image on the unit sphere S is finally

$$\lim_{B \to 0} B/S = (|X_u \wedge X_v|)/(K|X_u \wedge X_v|) = 1/K. \qquad (2.55)$$

For a detailed proof see DoCARMO (1966). The local density on the sphere is proportional to 1/K, or for a larger surface area we expect the density at a point p to be

$$\left[\frac{\iint K \, |X_u \wedge X_v| \, du \, dv}{\iint |X_u \wedge X_v| \, du \, dv} \right]^{-1} \qquad (2.56)$$

that is the inverse 'weighted average' of the Gaussian curvatures. There is one point one has to take care of. The Gaussian curvature K can assume positive and negative values -- thus, it is necessary that the Gaussian curvature does not change sign on the surface element under consideration. Otherwise we cross a point where K=0, and at that point the density is not defined (equation 2.56).

However, if K=0 everywhere at the surface, one has to distinguish two cases. If the surface is planar, then there exists no density distribution; all normals are mapped onto the identical point on the sphere. In the case of a parabolic surface element, e.g. a cylindrical or conical one, we can define densities if the Gaussian curvature K is replaced by the curvature k of the generating curve of the surface; the surface elements are replaced by arc length. The concept of a density distribution, therefore, is rather complex for orientation data from surfaces. One has to distinguish various cases, and one can handle only finite surface elements that satisfy certain conditions.

Anyway, it is as usually, the most interesting situations are those which cause trouble. In this case, interesting situations arise when the surface contains points at which the Gaussian curvature changes sign. Such situations arise in the most simple cases, e.g. if the surface is a sinusoidal cylinder -- in this case, the Gaussian curvature K changes sign at the inflection points of the generating sine function. Somewhat more complicated surfaces are given in Fig. 2.31. The lines of parabolic points, which divide the surfaces, appear also on the Gauss map (stereographic projection) where they bound the area of normal vectors. The density of normals on the sphere is especially high at (or near) the boundary of parabolic points which define a fold line of the Gauss map (cf. Fig. 2.2). The concept of folded maps applies also to the stereographic projection in the vicinity of parabolic points on a surface. The high (theoretically infinite) density

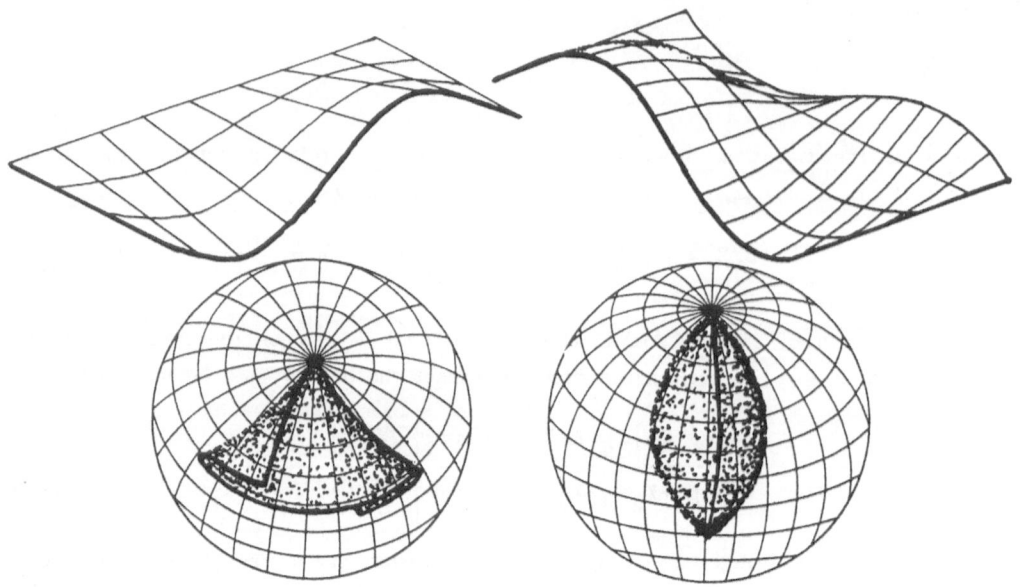

Fig. 2.31: Complex "sinusoidal" surfaces -- the sets of parabolic points on the surfaces map onto 'caustic lines' (with high densities) on the unit sphere. These 'caustics' are typical 'catastrophe singularities'.

of normals at these singularities has its analogy in caustics, and an interesting application to phonon focusing (cf. DANGELMAYER & ARMBRUSTER, 1983, for a thorough theoretical discussion). These patterns likely apply to the focal mechanism data in seismology with minor modifications. Further, such singularities occur at least in the theoretical simulation of plastic deformations (Fig. 2.32) as they were studied by LISTER et al. (1977). The singularities, which bound the distribution of normals are closely related to the singularities studied in catastrophe theory (DANGELMAYR & ARMBRUSTER, 1983), and some of them will be analyzed in more detail in the 4th chapter. In geology a common problem is to find the 'fold-axis' from a set of tangent planes. The linear approach is to find a "regression circle" under the assumption that the tectonical folds are cylindric (or conic). The approach by "caustic lines" probably allows to gather additional information as soon as the surfaces are more complicated.

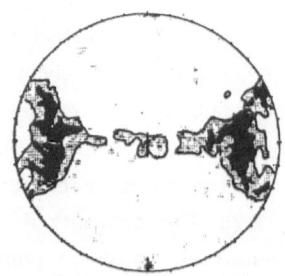

Fig. 2.32: c-axis pole figures for model quartzite under plane strain (modified from LISTER et al. 1978).

2.4 RECONSTRUCTION OF SURFACES FROM SCATTERED DATA

Contouring by computer became a widely used method in geology (e.g. DAVIS & McCULLAGH, 1975; KRUMBEIN & GRAYBILL, 1965; HARBAUGH et al., 1977; FREE-MAN & PILRAU, 1980). The use of computers, however, causes special problems which do either not occur within the classical hand method or are simply not recognized when contours are interpolated from a hand made triangulation net. Fig. 2.33 illustrates with a 'catastrophic' example how different computer results may be dependent on the estimation method. Fig. 2.34 gives a much earlier hand made estimation for the identical data set (BAYER, 1975).

The triangulation method (cf. CAVENDISH, 1974) is, in general, not used for computer procedures because it needs too much 'intuition' to create a triangulation net from scattered data points. On the other hand, it is relatively simple to interpolate and draw contours from a regular grid (e.g. SCHUMAKER, 1976). Therefore, contouring procedures for computers use mostly a two pass process:

** in a first step, the surface data are estimated for the points of a regular grid,

** in the second step, the contour lines are interpolated from the regular grid.

The final computation of the contour lines is a rather stable process. Useful methods are well known from finite elements (ZIENKIWITZ, 1975) and from spline interpolation. Even more complex methods like global or semilocal surface fitting can be used to derive contour lines from regular grids (SCHUMAKER, 1976). Anyway, on the level of the local grid cell the estimation of contour lines is not uniquely determined, an aspect which will be discussed in some detail.

The major problems, however, derive during the estimation of the grid values, and this will be the major theme of the following sections. The central example is an estimation method by minimal convex polygons, which is capable to illustrate the problems arising during the estimation of the grid values.

2.4.1 The Regular Grid

The main problem with scattered data is to evaluate the surface values for the regular grid. In geology, in particular, this process can be complicated because the data commonly show a natural ordering along outcrop lines such as river beds, tectonic zones etc. Therefore, very sophisticated methods have been developed for the gridding process. For instance, weighting functions are used that include directional searching of data points; also rather complicated statistical methods have been developed for the prior

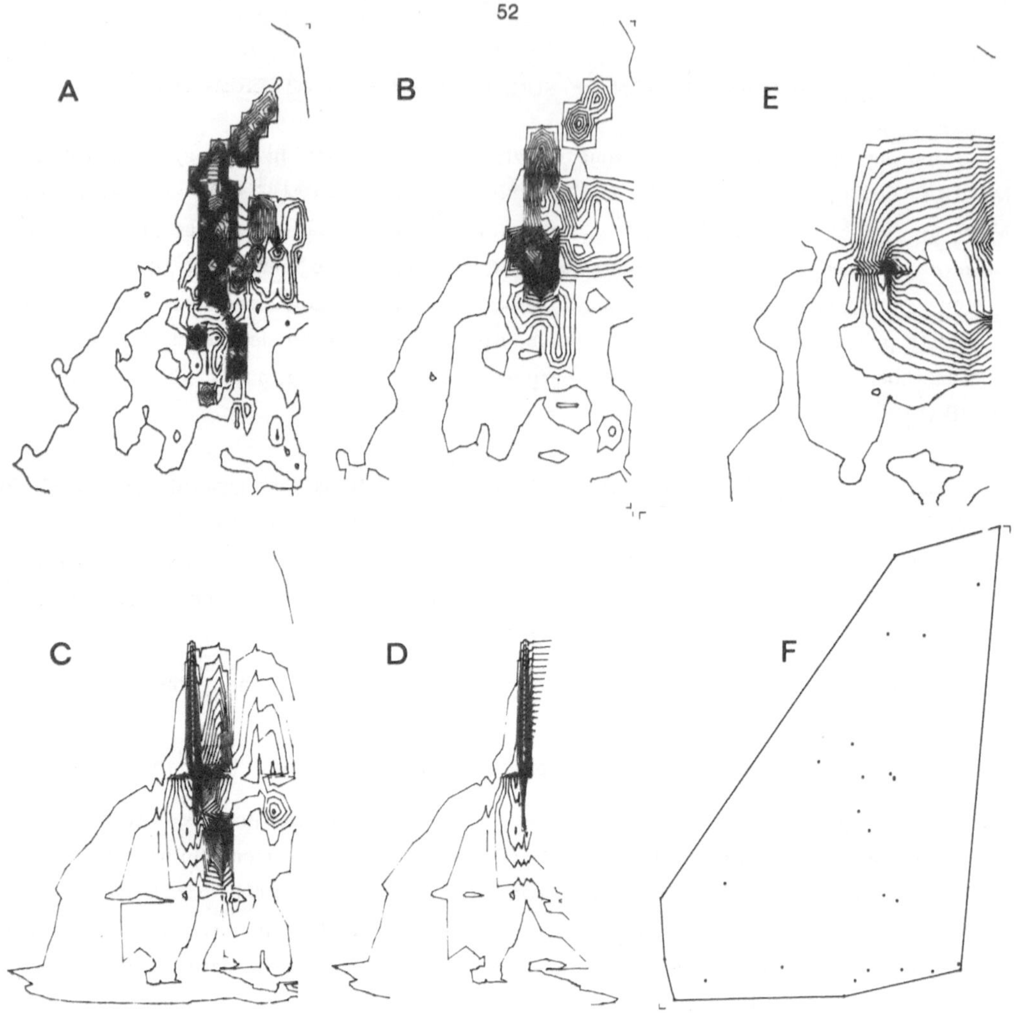

A   B   E

C   D   F

Fig. 2.33: 'Catastrophic' contour maps from a data set with strongly fluctuating surface values. a,b: regular grids of different roughness; minimal polygon method (see text). c,d: the same estimations on an irregular spaced rectangular grid, every data point is located on a grid point. e: grid point estimation by Shepard's weighting function (see section 2.3.3). f: point distribution and bounding polygon.

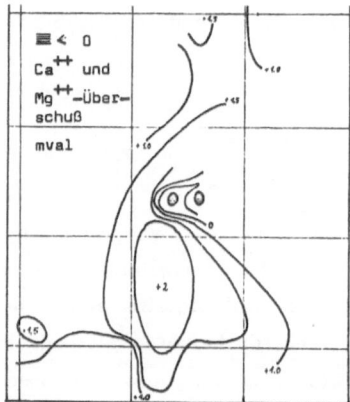

Fig. 2.34: A 'hand' estimation of contour lines for the data set of Fig. 2.33.

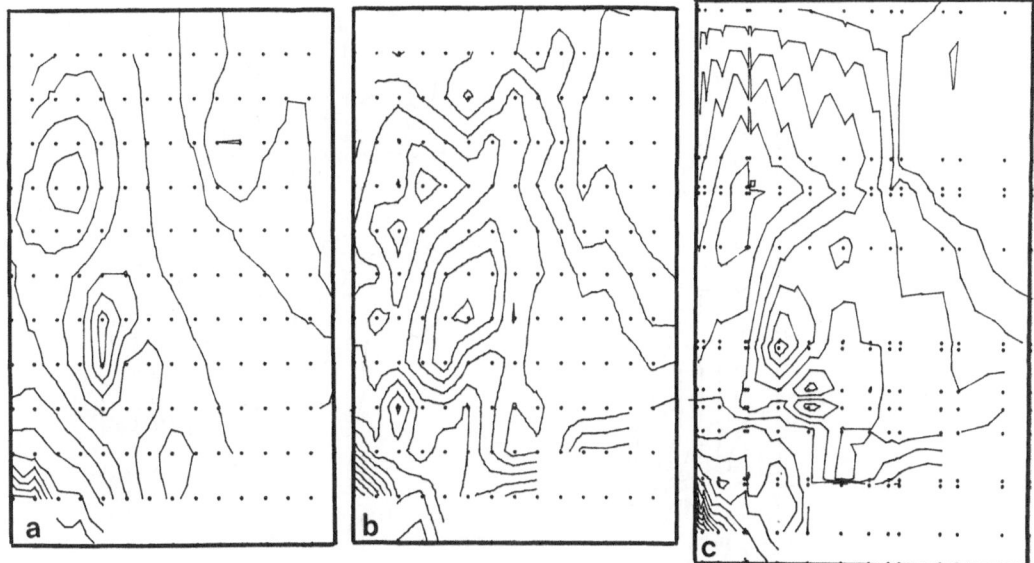

Fig. 2.35: Three possible solutions of isolines for the identical data set. a) Approximation by Shepard's method; b) minimal polygons; c) minimal polygons when the grid is generated from the data points. Grid points are marked. Modified from ALTHEIMER et al., 1982.

and posterior analysis of the grid values (JOURNEL & HUIJBREYTS, 1978; HARBAUGH et al., 1977; HUIJBREYTS, 1975).

In principle, the gridding technique is a map from one finite point set onto another finite point set, whereby no one to one correspondence exists. There may be more grid points than data points, or there could be less grid points than data points. This relationship may also change locally within the global gridding structure, as illustrated in Fig. 2.35 in terms of the grid points. Therefore, one may expect stability problems to occur, i.e. that the map becomes locally folded. Fig. 2.33 shows a 'catastrophic' result of computer contouring, due to the special, very inhomogeneous data configuration, the grid structure and the gridding methods. This configuration was chosen to illustrate the problems that arise from computer contouring. As far as stability problems are involved, they will be discussed below.

A minor problem in gridding is the special structure of the regular grid (Fig. 2.35). In general, the grid (or net) is chosen in such a way that the distances between the net points are constant; at least they are constant for every coordinate direction. Therefore, the resulting contours will depend on the initial decision about the grid structure (Fig. 2.35). If the grid is too rough, data points will be lost, or they are not correctly recorded. On the other hand, if the grid is fine enough to provide locally the required

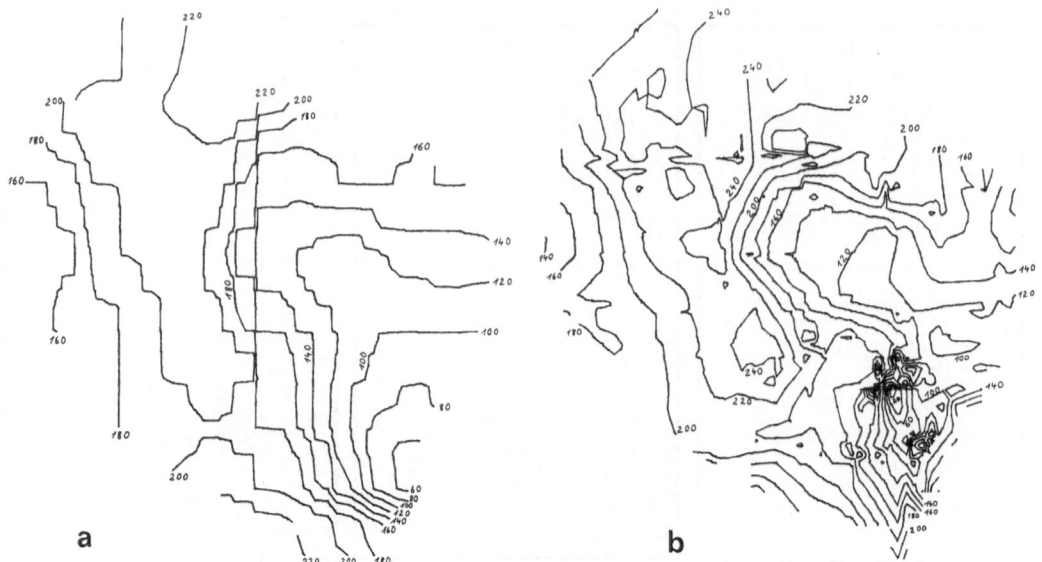

<u>Fig. 2.36:</u> Two isoline representations of foraminiferal diversity (entropy) in Todos Santos Bay, California. a) Shepard's method; b) minimal polygons. Adapted from ALTHEIMER et al., 1982; data from WALTON, 1955.

resolution, the net may be too fine in other areas where the data points are sparsely scattered (cf. Fig. 2.35 c). For a computer program the latter case may be more important than lost data points because a very fine grid can cause an extensive need of memory and unreasonably long computation times.

Furthermore, the original data points may not be recorded correctly, even on a very fine grid if they have intermediate position between the grid points. This can become important if the estimation procedure has a smoothing effect like it is the case with Shepard's local method (Fig. 2.36 a, cf. section 2.3.3). This problem can be solved, to some extent, if it is possible to choose the grid points in such a way that every data point becomes a grid point and that the grid is still a rectangular one (Fig. 2.35 c; ALTHEIMER et al., 1982). A really satisfying solution of this problem would require that the grid is formed over the data points. The classical way to do this is the triangulation method, but a good triangulation net is hard to establish within the computer (SCHUMAKER, 1976; CAVENDISH, 1974).

2.4.2 Global and Local Extrapolations

Another problem is to bound the estimated surface values to a reasonable area in order to avoid extrapolations into areas where no data points are available. Most gridding techniques allow such extrapolations, and most computer programs in use do not test this simple problem. As can easily be derived from the classical triangulation method (ALTHEIMER et al., 1982), the most extensive boundary giving reasonable estimates is the convex polygon that surrounds all data points, though it does not ensure a reasonable solution, as the Shepard estimation in Fig. 2.33 illustrates. Sometimes a better restriction of the solution space can be forced by additional constraints like a limit distance between data points and grid points. But such restrictions affect all estimations, not only the boundary -- and a typical result are 'holes' within the working area.

There are several possible strategies to establish a convex boundary for the data points (cf. ALTHEIMER et al., 1982). However, we shall see in the next section that a very simple method results from 'minimal polygons', which can be used to eliminate the interior points of the bounding polygon.

Indeed, the problem to find a reasonable global boundary for the working area has its local equivalent. One has to ensure that a grid point, for which a surface value is evaluated, is located within a closed polygon of data points and that these points are used to estimate the surface value. Otherwise, local extrapolations may occur, which cause maxima or minima, which are not represented in the data or 'platforms' result if the local estimation uses weighting functions.

The sectorial search method is an approach to solve this problem (HARBAUGH et al., 1977). The problem, of course, does not occur with the classical triangulation method because, in this special case, the whole area in question is densely covered by triangles or by convex polygons with the data points at the corners. Again, most of the gridding techniques in use do not necessarily test these conditions, especially the weighting average methods do not. For the usually complex geological data the

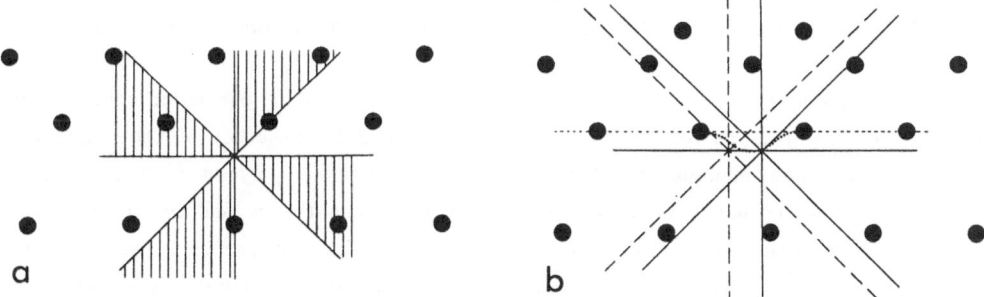

Fig. 2.37: a) the octant search method. A data point must be found in every octant for the estimation of a grid point value. This avoids local extrapolations. b) Surface reconstruction by an octand search method generates sinusoidal contours even from a simple cylindrical surface.

problem has been well recognized (HARBAUGH et al., 1977), and strategies like quadrant and octant search patterns have been developed to overcome the problem (Fig. 2.37). These methods require that for every grid point a set of data points must be found in a certain number of radial sectors (Fig. 2.37). Although the procedure secures that the grid point is surrounded by a -- not necessarily convex -- polygon of data points, it may produce other defects. A local gap will be produced if one sector is empty, even if there exists a convex polygon which includes the grid point and which would secure a useful local estimation. From this viewpoint, the method is not flexible enough because it requires locally a specific data configuration. On the other hand, the process tends to irregular contours because the surface values estimated from the moving sector system do not change smoothly, but they change rather suddenly when a data point enters or leaves the system (Fig. 2.37), at least, if the required number of data per sector is small. Conversely, a higher number of required data causes an increasing number of gaps within the solution area.

2.4.3 Linear Interpolation by Minimal Convex Polygons

The discussed problems caused us in 1980 to develop in the 'Sonderforschungsbereich 53, Palökologie' (ALTHEIMER et al., 1982) a gridding technique that satisfies all the so far discussed conditions in a very simple way -- the grid values are estimated from the minimal bounding polygon of a grid point, by a triangle. The method works rather well up to the point that sometimes strange local extrema appear in areas where no data points are available, i.e. the extrema cannot be explained by the data structure. A later stability analysis provided some remarkable results which also holds for other gridding techniques, especially for the discussed sector search methods and for the classical triangulation method. It will turn out that the problems are mainly geometrical ones and that our gridding technique is not a really good way to estimate surface points, but that it illuminates very clearly the problems of contouring from scattered data.

As was pointed out above, the locally stable estimation of grid point values requires that the grid point is located within a polygon spanned by a subset of the original data points. For a plane mapping problem the smallest possible polygon is a triangle with data points at its corners. As is well known from polygon theory and linear optimization (COLLATZ & WETTERLING, 1971), this minimal polygon has the property that the surface values can be simply estimated by a linear interpolation for any point within the triangle and along its boundaries (cf. ZIENKIEWITZ, 1975; SCHUMAKER, 1976). Geometrically the three points establish a plane surface element. Furthermore, the principle of 'nearest neighborhood' can be easily satisfied. 'Nearest neighborhood' means that a grid point value should be estimated from closest data points (as far as this does not cause local extrapolation). If there exists a polygon at all that encloses the grid point in question, then there exists a convex polygon, especially a triangle, which both includes the grid point and has the point of nearest neighborhood at one corner. The proof of this statement can be outlined in the following way:

> If the point of nearest neighborhood is connected with every point on the convex boundary of all data points, then the whole area inside the global convex boundary is densely covered by non-overlapping triangles, and the grid point must be either an interior point or a boundary point of one of these triangles.

This secures that we find a bounding triangle. The minimal triangle without an interior point is found with the following strategy:

> Start at the point of nearest neighborhood and find the second and third nearest points which form, together with the point of nearest neighborhood, a bounding triangle for the grid point.

We can call this triangle the minimal convex polygon -- minimal because these are points of minimal distance with respect to the convexity condition. One should expect that such a triangle is a locally optimal form for the approximation of the grid point value. The approximation turns into a local interpolation. In addition, the local properties of the triangles project onto the global problem to find the bounding polygon of all data points. The local triangulations are bounded to the interior of the convex polygon boundary. This gridding technique has, therefore, the additional advantage that it can be implemented in a very compressed way on a computer (ALTHEIMER et al., 1982) and that it provides automatically control over the boundaries of the working area. On the other hand, one can use the method to eliminate all points inside the global convex boundary of the data set. The points to be eliminated are simply those for which a minimal polygon exists which has data points on its corners -- or the global convex polygon consists of the data points which are not interior points of a triangle with data points at its corners.

2.4.4 Stability Problems with Minimal Convex Polygons

Having done all these analyses of local and global properties, it was surprising for us that the implemented process became unstable with certain data configurations (cf. Fig. 2.33). The main anomalies are local extrema which are not justified by the data. They occur mostly within relatively large areas without data points, and they are especially strong if the surface values of the nearby data points are very irregular.

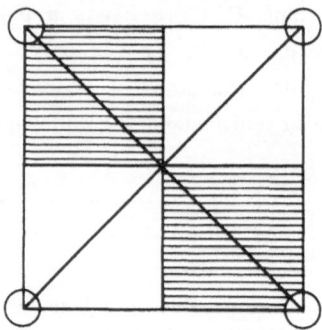

Fig. 2.38: The subdivision of a rectangle by the local triangulation method (minimal polygons). The blank areas belong to the same triangulation (as the shadowed do).

In addition, the anomalies are very sensitive to small changes of the grid structure. Some of these patterns resemble very closely the problems which may arise from sectorial search methods. To explain these anomalies it is necessary to study the local structure of the polygons, and to see how triangles may come into competition during the gridding process.

Four points form the minimal polygon, which can be further divided into triangles. For any grid point located within this polygon two possible triangulations exist (Fig. 2.38). Which triangle will be chosen, depends on the distance functions between the grid points and the polygon corners -- i.e. on the nearest neighborhood rule. The minimizing condition for the distance function divides the polygon into four regions (Fig. 2.38). The opposite rectangles belong, thereby, to the same triangulation, and, therefore, have stable and nearly smooth solutions in their interior. But the two possible triangulations give different solutions which are not connected in a smooth way (Fig. 2.39). If the grid point moves over the triangulation boundaries, then the solution jumps from one interpolation surface onto the alternative solution. The instabilities, therefore, occur

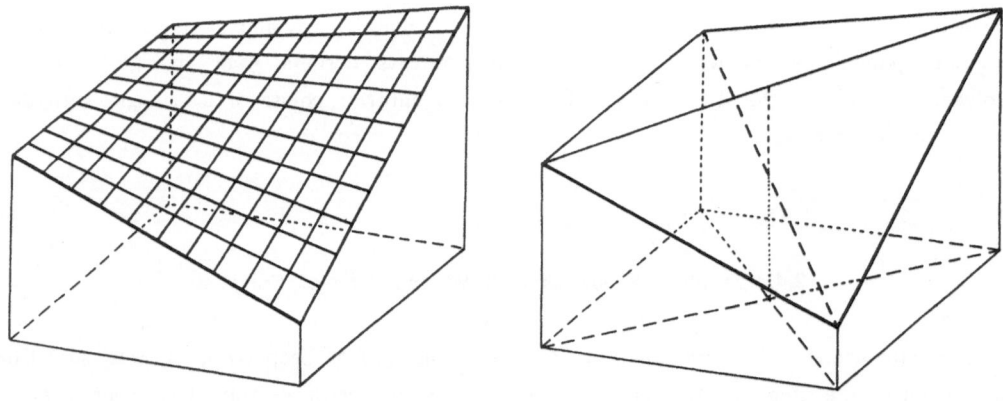

Fig. 2.39: A ruled surface over a rectangular grid element and its approximation by the two possible triangulations which yield different solutions.

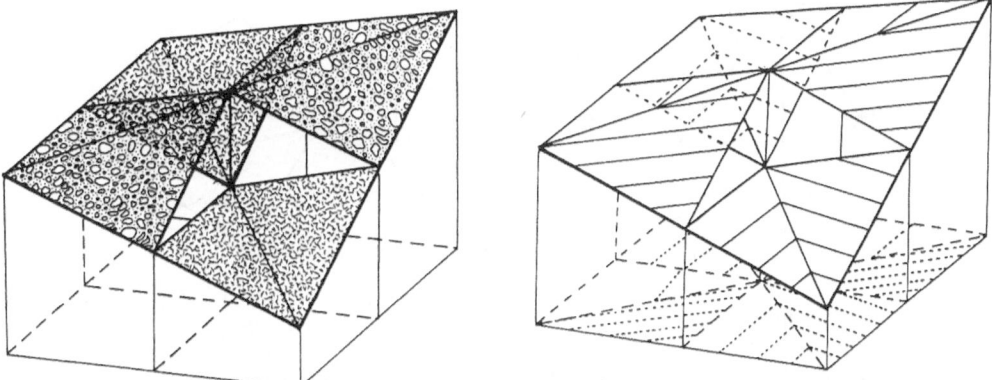

<u>Fig. 2.40:</u> The bifurcation surface which corresponds to the two possible triangulations of Fig. 2.39 (cf. Fig. 2.38). The instability is independent of the density of the grid, it results only from the local triangulation method. On a sufficiently fine grid the contours reflect the bifurcation of the surface (b).

in two ways:

1) Closely related grid points may deviate strongly in the estimated surface values because they belong to different local triangulations.

2) Even small changes in the structure of the regular grid can locally cause rather dramatic changes in the estimated surface values -- e.g. for two independent contouring processes over the identical data set.

These effects are especially strong within relatively large areas without data points, i.e. whenever the approximation involves data points which are far apart.

Fig. 2.40 illustrates in more detail how the local solution over a rectangle bifurcates into two disconnected surfaces with discontinuous contours. From which interpolation surface the estimated value will be taken, depends, as discussed, only on the position of the grid point(s). The situation becomes even worse if one considers higher polygons, i.e. a larger number of data points in competition. The possible number of local triangulations increases rapidly with the number of polygon corners (Fig. 2.41). The approximation process will become more and more instable as the number of data points in competition increases -- a situation favored by large empty areas within the data space. A curious situation is that in such cases a refinement of the grid increases the instability in the way that the resulting surface approximates the local discontinuities of the triangulation rather than a continuous surface (cf. Fig. 2.40 b). The situation is nearly the same with the octant search method where the triangles are replaced by open angular sectors.

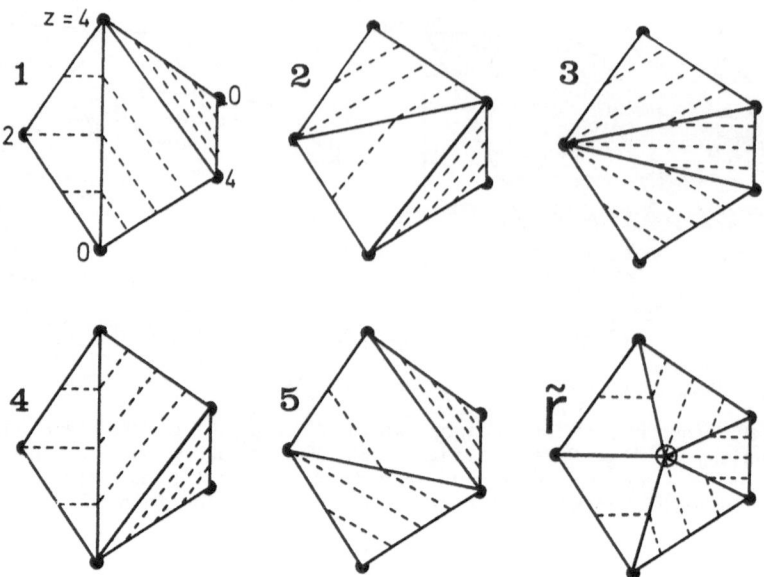

Fig. 2.41: The various possibilities for the triangulation and linear surface approximation of a five-point polygon (1 to 5: the numbers z=1 ... indicate the height of the corner points; r: the centered interpolation -- the central point is the arithmetic mean of the corner points.

One could argue that the discussed instabilities are only problems of the local approximation method. But if we transfer the results to the classical triangulation method, which covers the data space densely with triangles, then it is not hard to see that we have, in principle, the same problems. Let several people draw a map from the identical data set with a free choice of triangulation, then the results can diverge to a large extent. The different solutions for a five-point polygon in Fig. 2.41 can be taken as an example. What we find is that the instabilities appear now exclusively between different maps.

We can go a step further. The same problems encounter us again when we draw contour lines from the estimated regular grid. A simple way to do this is again a triangulation of the grid -- the contour lines are uniquely determined on a triangle. However, there are several possibilities for this triangulation, some of them are illustrated in Fig. 2.42. Every possible triangulation gives another solution, and we can only hope that these solutions deviate not too widely because the surface is simple enough. Besides this triangulation method other strategies are in use which draw the contour lines directly from the rectangles, e.g. such a strategy is used in the program package SURFACE II (SAMPSON, 1975). Within this strategy problems arise when two opposite corners of a cell are higher and the two others are lower than the value for the contour line enter-

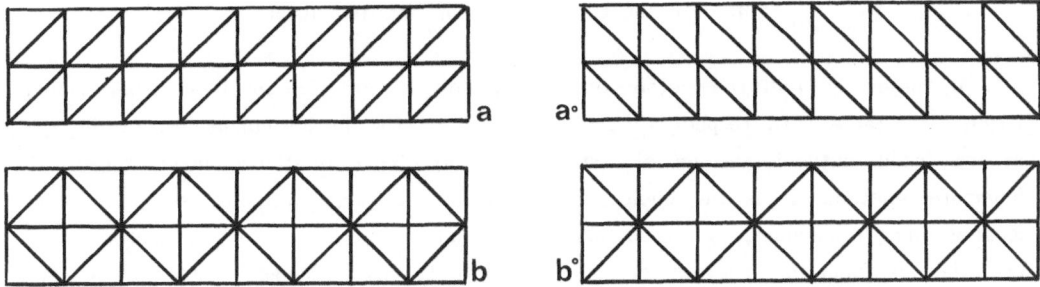

<u>Fig. 2.42:</u> Four alternative triangulations of a regular grid.

ing the cell. Fig. 2.43 illustrates this situation, and it becomes immediately clear that the cases (b) and (c) are just the two alternative triangulations of the grid cell, the decision problem is the same as discussed earlier. Case (a) is slightly different, it represents a centered grid element where the central value could have been estimated as the arithmetic mean of the corners (cf. Fig. 2.41). This approximation will be dis-

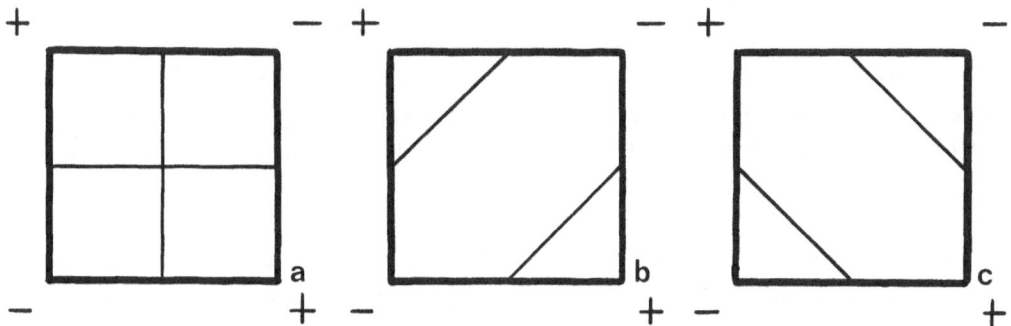

<u>Fig. 2.43:</u> Possible paths of contour lines through a grid cell -- (+) corners higher and (-) corners lower than average of corner values. Modified from SAMPSON (1975).

cussed in the next section. In the SURFACE II program a decision is made between the solutions (b) and (c) in Fig. 2.43: (b) is chosen if the average of corner values is higher than the entering contour line while (c) is chosen if the average is lower than the value of the contour line. This choice is arbitrary, however, it ensures that contour lines do not intersect within the grid element (Fig. 2.43 a) -- this switch causes a jump from the lower to the upper surfaces in Fig. 2.40 when the average height of the corner points is passed.

2.4.5 Continuation of a Local Approximation

It turned out that the method of minimal polygons or of a local or global triangulation is instable with respect to small changes of the initial conditions. In the case of the 'hand method', the initial condition is the choice of the triangulation, in the 'computer method', it is the choice of the grid. The same problem extends to other local gridding techniques, to the sectorial search methods and even to the approximations by weighting functions. They all are very sensitive to small changes of the initial conditions and to changes of the parameter setting. A major problem arises if there are large areas without data points. In this case, the interpolation process can be somewhat stabilized if one does not use the minimal convex polygons but tries to find the locally maximal convex polygon. However, competition between polygons can be only avoided if the entire interior of a locally bounded polygon is treated as a local continuum, and if all grid points inside the polygon are estimated from its corner points by some smooth process. The competition during the formation of local polygons can be avoided if the local solution projects continuously into the neighborhood, i.e. no overlapping polygons are allowed until they have the same solution inside the intersecting areas and on the common boundaries. The problem has a formal analogy in the analytic continuation of a function in the complex plane. This analogy suggests that one could start from a local solution, a local contour line, and then construct its continuation through the data space by use of some convergence criteria. The convergence circle of the analytic problem could thereby be replaced by convex polygons over the finite data set.

The previous remarks lead to a geometrical problem, which is hard to solve in the case of randomly scattered data. Nevertheless, it seems useful to discuss finally how the linear interpolation over triangles can be generalized for any convex polygon and how a local solution over a regular grid element can be extended throughout the global data space.

A) A Local Continuous Approximation

In the case of a rectangle, the simplest approach toward a stable continuous surface approximation is to construct a bilinear function over the corner points (SCHUMAKER, 1976)

$$f(x,y) = a_1 + a_2 x + a_3 y + a_4 xy. \qquad (2.57)$$

The corner values of the grid element have to be used to determine the coefficients. Now, any rectangle can be standardized to a square of unit area by the map

63

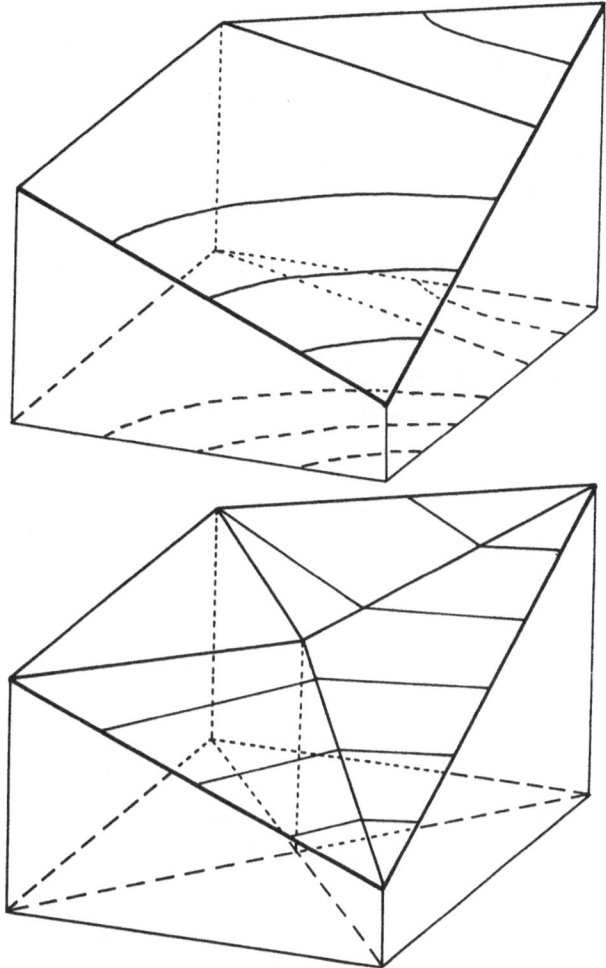

Fig. 2.44: Surface interpolation over a rectangle by use of a bilinear function (above); for details see text. The bilinear interpolation can be approximated by a linear interpolation if an additional central point is used, which can be computed as the arithmetic mean of the corner points.

$$x \dashrightarrow (X_{i+1} - x)/(X_{i+1} - X_i)$$

$$y \dashrightarrow (Y_{i+1} - y)/(Y_{i+1} - Y_i),$$

(2.58)

where X_i and Y_i are the coordinates of the corner points.

Besides standardization, the map (2.58) transforms the global grid coordinates into local ones. Using the new local coordinates, the linear interpolation along the boundaries of the rectangle can be expressed as

$$f(x) = w_2 F(x=0) + w_1 F(x=1),\qquad\qquad (2.59)$$

where the F-values are the surface height at the corner points and the w_i are weighting functions: $w_1 = x$, $w_2 = 1-x$, x in local coordinates (for the y direction x has to be replaced by y).

The approach by a bilinear function implies to construct a two-dimensional weighting function from the product $w(x,y)=w(x)w(y)$ (e.g. PFALTZ, 1975; DeBOOR, 1978). If the weighting functions for the boundaries are inserted, one finds

$$
\begin{aligned}
w_1(x=0,y=0) &= (1-x)(1-y) &= w(1-x,1-y) \\
w_2(x=0,y=1) &= (1-x)y &= w(1-x,\ y\ ) \\
w_3(x=1,y=1) &= xy &= w(\ x\ ,1-y) \\
w_4(x=1,y=0) &= x(1-y) &= w(\ x\ ,1-y)
\end{aligned}
\qquad (2.60)
$$

and
$$w(x,y)=xy,$$

a very simple pattern of permutations of the coordinates, which easily can be pro-- grammed. It is easy to prove that $\sum w_i = 1$, and that $z(x,y) = \sum F_i w_i(x,y)$ is just the earlier noticed bilinear function which provides a continuous surface approximation over the grid element. The weighting functions have the property that

$$\sum F_i w_i(0.5,0.5) = \sum F_i/4,\qquad\qquad (2.61)$$

i.e. there exists one point on the surface which is simply the arithmetic mean of the corner points. This observation allows a first order approximation of the bilinear surface over a rectangle by a simple triangulation. If one adds the centroid of the corner points to the data points, then there exists locally a unique triangulation of the grid element which is given by the connections of the central point with the corner points. The surface estimated from this triangulation is a linear approximation of the surface, which was defined by the bilinear equation (2.57). Figs. 2.41 and 2.44 provide examples for this approximation. It is easy to see that a unique triangulation and, therefore, a unique local surface approximation can be constructed for any convex polygon with n corners. The additional central point is given by

$$(x_c, y_c, z_c) = (1/n) \sum (X_i, Y_i, Z_i).\qquad\qquad (2.62)$$

It may be useful to introduce a meaning for this interpolation scheme. The bilinear model is a harmonic function, and this allows a physical interpretation. If a sheet of rubber is stretched over the rectangular boundary, the resulting surface equals the surface described by the bilinear equation. For the generalized convex polygon with n corners one can construct such a surface in the following way (BETZ, 1948): The convex polygon is mapped onto the unit circle by means of the Schwarz-Christophel formula. The boundary values are then evaluated in terms of a Fourier series. The required harmonic function on the unit circle is finally expressed by the equation $f(r,\psi)=a_0/2+ \sum r^n(a_n\cos(n\psi) + b_n\sin(n\psi))$, and a first approximation on the original polygon is given again by the centered triangulation.

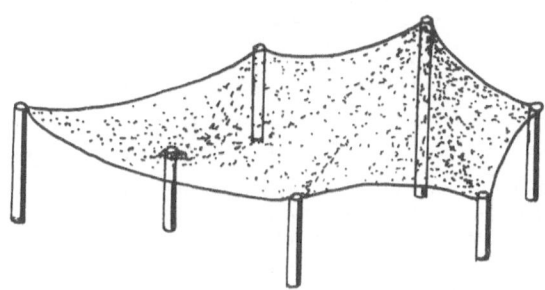

Fig. 2.45: A tent structure provides an example of a continuous surface with discontinuities at the poles.

B) Continuation of a Local Surface Approximation

The centered grid element, as defined above, leads in a rather natural way to a continuous solution over the global regular grid structure. If we add the computed central grid points to the grid, we have simply a refinement of the grid, and we can repeat this process infinitely. In the second iteration, additional grid points and values are computed at the boundaries between the original grid elements and provide a continuous approximation between grid elements. In general terms, a regular approximation within grid elements occurs at every odd interpolation step while at even steps overlapping grid elements are continuously connected. What we find, is a surface which everywhere satisfies the Laplace equation

$$u_{xx} + u_{yy} = 0, \qquad\qquad (2.63)$$

a surface, which is everywhere smooth, only at the grid points local discontinuities

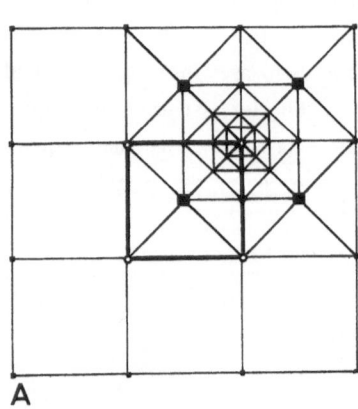
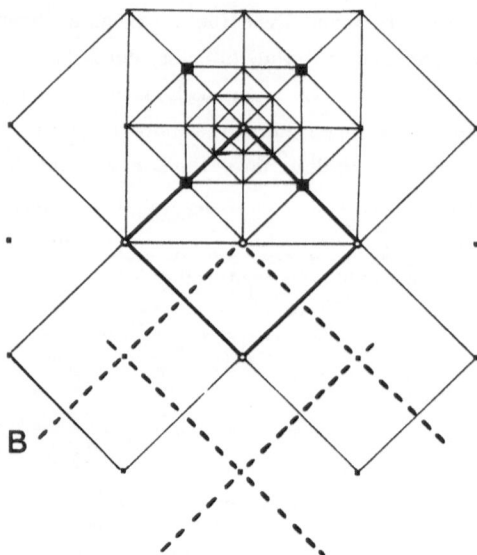

Fig. 2.46: Iterative refinement of the grid structure by recursive averaging. Only a single pathway is illustrated, which asymptotically approaches a corner of the central grid element. Any other point, the original grid points and the averaged ones, cause similar cascades: The original centroid grid element is subdivided into smaller and smaller rectangles which, in the limit, cover the area densely, however without being continuous in a differentiable sense. Left: a regular orthogonal grid, right: a centered regular grid which actually consists of two overlapping grids as indicated.

appear (Fig. 2.45). The relation between the recursive averaging process and the Laplace equation can easily be shown if the Laplace equation is approximated on a finite grid. In terms of a finite grid, equation (2.63) reads

$$(U_{i-1,j} - 2U_{i,j} + U_{i+1,j}) + (U_{i,j-1} - 2U_{i,j} + U_{i,j+1}) = 0 \qquad (2.64)$$

providing a finite approximation, which can be rewritten as

$$U_{i,j} = (1/4)((U_{i-1,j} + U_{i+1,j}) + (U_{i,j-1} + U_{i,j+1})), \qquad (2.65)$$

and this is simply the average discussed above. Thus, our continuation process is a finite analogue to the analytic continuation in the complex plane.

The continuation process by recursive averaging is, in addition, optimal in the sense that it is stable under small disturbances of the grid pattern and is optimal in terms of computation costs. To see, why the latter remark holds, let change the viewpoint again to a single grid element. We want to find a local approximation which is continuously connected with the neighborhood. To find such a local approximation we need in totality 16 grid points like in Fig. 2.46a, and this elementary grid allows refinement to any level within the central grid element. An alternative would be to use initially a centered grid (Fig. 2.46b) which, of course, provides an initial triangulation. However, if we request a continuous connection with neighboring elements, we need 21 grid points. The increased number of necessary grid points can be related to the fact that the centered grid is not unique, i.e. that there exist two alternative grid structures as indicated in Fig. 2.46b. A continuous solution requires that these alternative grids are superimposed, and this causes the higher number of required grid points.

However, the continuation problem can be solved in a quite different way: We can request that the local surface element has continuous derivatives along its boundaries and, thus, can continuously be connected with the neighboring elements. Such an approximation requires at least cubic splines, and first we consider the case that the first derivative vanishes along the boundaries of the grid element. A useful approximation is given by the weighting function

$$w(x,y) = w(x)w(y) \tag{2.66}$$

and $\quad w(x) = x^2(3-2x); \quad w(y) = y^2(3-2y).$

The height of a surface point can be expressed as weighted average of the height of corner points

$$z(x,y) = Z_1 w(x,y) + Z_2 w(1-x,y) + Z_3 w(1-x,1-y) + Z_4(x,1-y). \tag{2.67}$$

If we use equations (2.66), we can rewrite equation (2.67) as

$$z(x,y) = ((Z_1+Z_4)-(Z_2+Z_3))(x^2(3-2x))(y^2(3-2y)) + (Z_3-Z_4)(x^2(3-2x)) + (Z_2-Z_4)(y^2(3-2y)), \tag{2.68}$$

an equation which looks rather complicated. However, if we introduce the abbreviations

$$u = x^2(3-2x); \quad v = y^2(3-2y), \tag{2.69}$$

equation (2.68) turns into a simple bilinear equation

$$z(x,y) = auv + bu + cv \tag{2.70}$$

with obvious parameter identifications for 'a', 'b', and 'c'. Thus, we are still dealing with equation (2.57), with the only difference that the coordinates (x,y) are replaced by functions of these coordinates. Equations (2.70) and (2.69) provide a system of equations consisting of two parts: The interpolation equation, which is simply a bilinear equation, and a map (x,y) → (u,v), which defines a deformation of the original coordi -

es, thus, that they satisfy certain conditions at the boundaries of the grid element. Our approximation problem turns into the problem to find a proper map $(x,y) \rightarrow (u,v)$ which satisfies the required conditions: The map

$$
\begin{aligned}
u &= a_1 x^3 + b_1 x^2 + c_1 x + d_1 \\
v &= a_2 y^3 + b_2 y^2 + c_2 y + d_2
\end{aligned}
\qquad (2.71)
$$

for instance allows to adjust the first derivative along the boundaries of the grid element; however, it is not the most general case (for a discussion of splines, see DeBOOR, 1978). Anyway, even an approximation by equation 2.71 is rather sensitive to small changes in the grid structure. The discussion of the cubic spline in section 2.3.3 can easily be extended to the two-dimensional case -- a small disturbance of the estimated slope at the grid boundaries can totally change the estimated surface pattern, which may switch from a ridge to a valley or vice versa. As easily can be seen, the stability of our approximation problem depends only on the map (2.69), and this relates it to singularity theory. We can transform the map (2.69) to a more convenient forms by a simple

dislocation of the origin $x \rightarrow x + 1/2$; $\quad y \rightarrow y + 1/2$ and $u \rightarrow u - 1/2$, $\quad v \rightarrow v - 1/2$, which yields

$$
u = 2x^3 - (3/2)x; \qquad v = 2y^3 - (3/2)y,
$$

a rotation $x \rightarrow x + y$; $\quad y \rightarrow x - y$ yields

$$
\begin{aligned}
u &= x^3 + 3x^2 y + 3y^2 x + y^3 \\
v &= x^3 - 3x^2 y + 3xy^2 - y^3
\end{aligned}
$$

and a final rotation in the (u,v)-space $\quad u \rightarrow u+v$, $\quad v \rightarrow u-v$ transforms our original map into

$$
u = 2x^3 + 6xy^2; \qquad v = 2y^3 + 6x^2 y, \qquad (2.72)
$$

a map which represents a special form of the double cusp catastrophe. In catastrophe theory such a map is embedded in a potential, in this case the potential would be

$$
V = x^4/2 + y^4/2 + 3x^2 y^2 - ux - vy, \qquad (2.73)
$$

and the map results from the condition that the partial derivatives vanish, i.e. from the equations

$$
\begin{aligned}
V_x &= 0 = 2x^3 + 6xy^2 - u \\
V_y &= 0 = 2y^3 + 6x^2 y - v .
\end{aligned}
$$

If we now return to the more general case (the derivatives are determined from the data points), we need again additional parameters. Catastrophe theory implies that the potential (2.73) has general unfolding

$$V = x^4 + y^4 + ax^2y^2 + bx^2y + cy^2x + dx^2 + exy + fy^2 - ux - vy, \qquad (2.74)$$

an expression which provides us with 6 free parameters to adjust the boundary conditions. This expression, however, is in local coordinates; in global coordinates we would have to unfold the map (2.72) and to consider all possible parameters inclusively the constant ones.

The Double Cusp is extremely unstable, the stable regions are extremely narrow, and even small disturbances cause switching solutions, in this special case switches from ridges and hills to valleys and depressions. This gives us a direct relationship to catastrophe theory; however, the problems encountered through these sections are connected with catastrophe theory in a much wider sense: The approximation problem in surface reconstruction is usually associated by some optimizing problem, i.e. to estimate the grid point from the nearest data points. A common problem with such optimizing strategies is that during a smooth change of the distance function 'the optimum solution changes with a jump, transferring from one competing maximum to the other' (ARNOLD, 1984). That, of course, is what we observed throughout the discussion of surface reconstruction. The connection of the observed instabilities with catastrophe theory may not necessarily be obvious because we usually think about discrete and non-differentiable systems in approximation processes. However, discrete are only the grid points, which turned out to be parameters of a smooth interpolation surface. By a smooth change of these parameters the approximation may react with sudden jumps in the geometrical solution. A change of the grid point values, however, is equivalent to a change of the boundary conditions, and these clearly affect the optimizing function. In terms of variable boundary conditions we can, therefore, apply catastrophe theory to the surface approximation problem. A discussion of more general optimizing problems is given in ARNOLD (1984).

Table 2.1 Singularities in optimization problems

Normal forms of a maxima function F (ARNOLD, 1984):

one parameter   two parameters

$F(y) = |\, y\, |$

$$F(y) = \begin{cases} |y| \\ \text{or} \\ \max(y_1,\ y_2,\ y_1 + y_2) \\ \text{or} \\ \max_x(-x^4 + y_1 x^2 + y_2 x) \end{cases}$$

3. NEARLY CHAOTIC BEHAVIOR ON FINITE POINT SETS

Chaos implies totally and apparently irremediable lack of organization. In physics, a classical example for chaos is turbulence. In a turbulent system, the pathway of a particle cannot be predicted at all, and two particles, which are initially close together, may depart in a short time interval. The transition from a deterministic (laminar) behavior to chaos (turbulence) can be usually described by a bifurcation tree (Fig. 3.1). "After the first bifurcation the flow becomes periodic, after the second bifurcation the flow is quasi periodic with two periods, and so on" (RICHTMYER, 1981). After a sufficiently high number of bifurcations the chaotic aspect of the flow is so highly developed that statistical methods are the proper way to study its behavior. It is clear that the behavior of such systems during the course of time depends very sensitively on the initial conditions (HAKEN, 1981), and that the bifurcations are not a dynamical feature, but appear in the state space of the system, i.e. they are a topological property of the system.

During the last decade, another way to study chaos has attracted much attention: the behavior of difference equations in calculators (MAY, 1974; RÖSSLER, 1979; THOMPSON, 1982). In this case, the dynamical system is replaced by an iterated map describing the outcomes in finite time intervals. MAY's (1974) favorite example was the standardized form of the logistic difference equation. A short review of the behavior of this equation will be given in the first section to introduce the concepts of bifurcations and of chaos more precisely. The explicit numerical approximation of a partial differential equation then elucidates once more the concept of bifurcation, and the concept of iterated maps is used to study infinite sequences of caustics in refraction seismics.

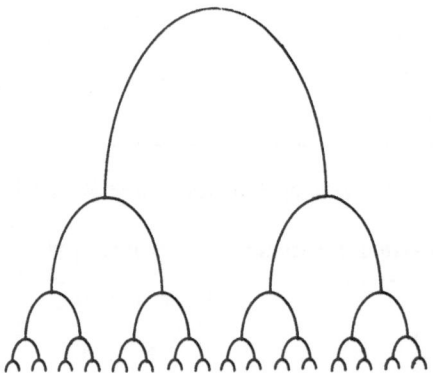

Fig. 3.1: A bifurcation cascade or a generalized catastrophe (THOM, 1975), as it results e.g. from the logistic difference equation.

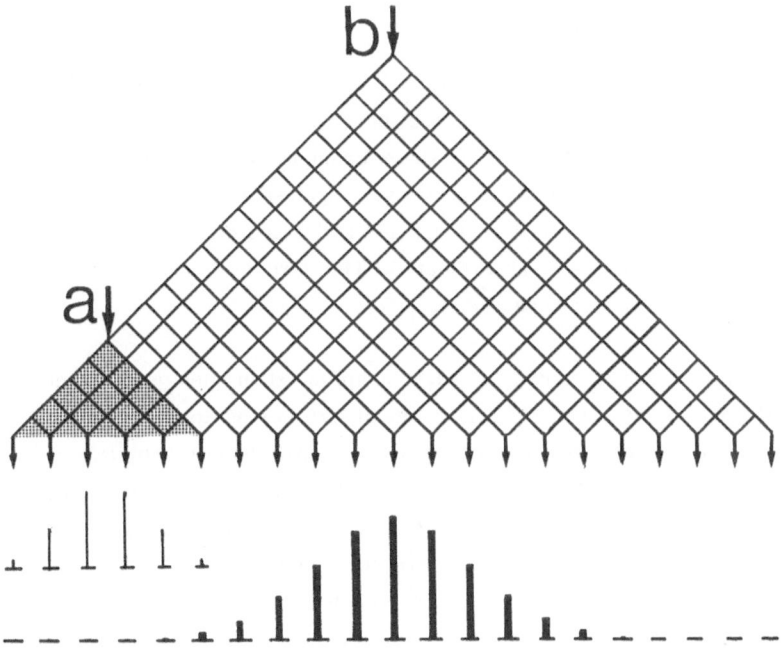

Fig. 3.2: Two versions of Galton's machine -- a small and a large one -- which produce the binomial distribution.

As was noticed above, statistical methods are the usual way to study chaotic systems. The 'Galton machine' (Fig. 3.2) illustrates the relationship between the chaotic trajectories of particles, which cannot be predicted, and the well predictable outcome if enough particles are considered. In this case, our impression of chaotic motion within the machine will not at least depend on its size (Fig. 3.2). In addition, the form, the internal geometry of the machine, affects the type of the statistical outcome. The bifurcation tree of Fig. 3.1 can be taken as another machine of this type. It will produce a uniform distribution. An interesting case occurs if the internal configuration of such machines depends on some parameters, or if the initial conditions can affect the outcome of the machine.

The first example is a brief review of the logistic difference equation. A more interesting example, from the geological viewpoint, is the instability of the explicit approximation of a partial differential equation. The bifurcation, which is caused by a smooth change of a parameter, can be nicely visualized by the uncoupling of the grid into two independent substructures. The concept of iterated maps is finally applied to series of caustics in refraction seismology.

The concept of bifurcations and chaos is then applied to several computer methods. The problem is that chaos in such cases is not obvious. In most examples, a small change of parameters will strongly influence the outcome, but with a computer procedure this sensitivity will normally not be detected because the data are only processed with a certain parameter setting. The first of these examples is the usual Chi^2-testing of directional data. The test is commonly performed against a uniform distribution, and it is unstable with respect to an arbitrary choice of the sectorial pattern on which the computation of the test statistic is evaluated. The striking point is that the stability of the test decreases with increasing sample size.

In the third section, problems with sampling strategies in sedimentology are discussed. One goal of the statistical analysis of profiles is to detect periodicity patterns. Two methods are in use, the analysis versus transitional probabilities and the classical time-series analysis. In both cases it is a typical strategy to take samples at equal distances. In this case, the transition matrix becomes dominated by singular loops, and the so-called 'transitional probabilities' are not further free of dimensions. In the case of a time series analysis, the identical approach can cause artificial pattern formation. The example is closely related to the genericity problem of maps, an aspect which is briefly mentioned. The main result will be that geometrical and geological reasoning cannot be replaced by a formal, pseudo-objective sampling strategy.

Then, we shall deal with various aspects of classical centroid cluster strategies. Again a situation is encountered where an increase of the sample size does destabilize a 'statistical' pattern recognition process, and it will turn out that these methods provide excellent examples of chaotic behavior on finite point sets -- they show the discussed properties of chaos, especially the extremely high sensitivity to small changes in the initial data.

Finally, the bifurcation of tree-like bodies is analyzed. The basic model is entirely deterministic; nevertheless the bifurcation patterns generated are rather chaotic. From this chaotic pattern, however, a well determined shape arises -- an analogy found in the shape of trees, which is typical on the species level. The analysis is based on a modification of HONDA's (1971) computer model and takes up the geometrical analysis, which roots in D'Arcy Thompson's and even Leonardo da Vinci's work.

3.1   ITERATED MAPS

Classically, stability is the most important concept for the numerical solution of differential equations. The typical way to solve differential and partial differential equations numerically is to transform them into an 'iterated map' by use of Taylor's theorem.

It is well known that there are sometimes several choices for the transformation, and that the various possible approximations behave differently with respect to the quality of the approximation, to the convergence and to other stability problems. Here some aspects of iterated maps are briefly discussed under topological aspects because this approach may give some insight not only in those problems, which occur with difference equations, but also in the concepts of bifurcations and chaos.

3.1.1 The Logistic Difference Equation

The logistic growth function plays some role in biology and in paleontology. The difference formulation of this equation was MAY's (1975) favored example for bifurcations and chaotic behavior. In the meantime, it became an important example for bifurcation cascades and chaos in various fields (e.g. HAKEN, ed. 1982). The differential equation of the logistic equation is given by

$$y' = ay(b-y), \tag{3.1}$$

which has a well known explicit solution. A simple straight forward difference approximation is given by

$$y_{i+1} = y_i + \Delta t a y_i(b-y_i). \tag{3.2}$$

For a special parameter setting of 'a' and 'b', the solution of this difference equation depends only on the parameter Δt, which represents a finite time interval. As Fig. 3.3 shows, the upper boundary is only approached for small values of Δt. As this parameter increases, one finds that the solution fluctuates around the saturation level. For larger values of the discrete time intervals, the long time output of the difference system resembles much more the Lotka-Voltera model (LOTKA, 1956) of a predator-prey system than the original logistic growth model.

By some elementary coordinate transformations (e.g. RÖSSLER, 1979) the logistic difference equation can be standardized to the form

$$y_{i+1} = ry_i(1-y_i), \tag{3.3}$$

which allows to analyze the behavior of this model in a general way. The relationship between the y_{i+1} and the y_i values can be plotted as the graph of a function for which the y_i values are the values of the independent variable. For the first iteration, the graph of this function is a parabola (Fig. 3.4, it I). The saturation value is exactly reached if $y_{i+1} = y_i$, and this defines a straight line in the (y_i, y_{i+1}) coordinates. In the

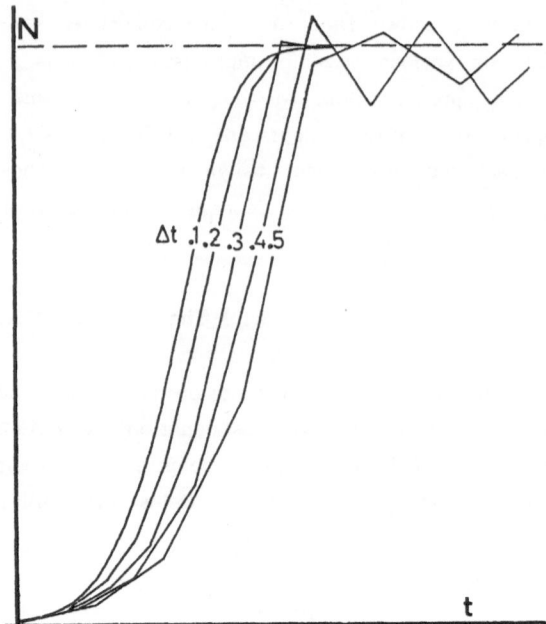

Fig. 3.3: Numeric solutions of the logistic differential equation for various discrete time intervals.

graph of function (3.3) the saturation point is given by the intersection of this line with a specific parabola, which is determined by the parameter r. Fig. 3.4 (I) shows how one can use these properties to analyze which values of r allow for a stable solution. The equivalent algebraic expression would be

$$y_i = r y_i (1 - y_i),$$

$$(3.4)$$

which can be solved for y_i. Higher iterations are capable to produce periodic solutions.

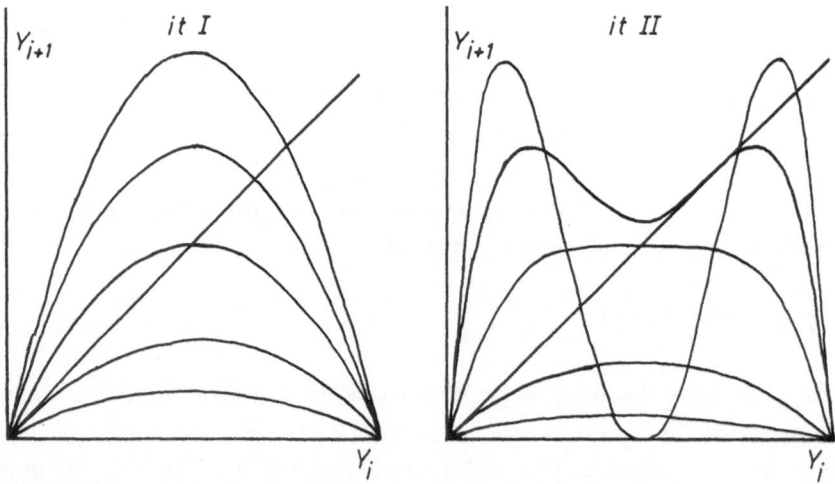

Fig. 3.4: First (I) and second (II) iteration system of the logistic equation in standardized form. The curves correspond to different values of the parameter 'r'; their intersections with the straight line are the equilibrium values.

The first one occurs for $y_{i+2} = y_i$, i.e. every second iteration takes the same value. Again one can find a graphic representation as well as an algebraic one. If one rewrites equation (3.4) in terms of y_{i+2} and of y_i, one finds a quartic polynomial

$$
\begin{aligned}
y_{i+2} &= ry_{i+1}(1-y_{i+1}) \\
&= (ry_i(1-y_i))(1-ry_i(1-y_i)).
\end{aligned} \tag{3.5}
$$

The stable points are found in the same way as before (Fig. 3.4, it II) by setting $y_{i+2}=y_i$, and we find up to four equilibrium points, but not all of them are stable. As the parameter r varies, one finds up to three intersections between the polynomial and the equilibrium line (except the trivial solution $y_i = y_{i+2} = 0$). In the same way we find an increasing number of periodic solutions for every relationship $y_{i+k} = y_i$, or, as k increases, we get an infinite number of periodic solutions or a bifurcation cascade like in Fig. 3.1. This type of chaotic behavior was analyzed by MAY (1974), who showed that the logistic equation has an infinite number of possible periodic trajectories and is of chaotic behavior.

However, the logistic equation is only the special celebrated example. OSTER & GUGGENHEIMER (1976) showed that any convex function can replace the parabola in equation (3.3) and drew connections to the Hopf bifurcation. Even a linear spline approximation causes such bifurcations and periodic solutions (Fig. 3.5a). Probably models based on exponential functions are more biological than the finite logistic model because they have no sharp upper limit. In the case of the finite logistic equation, there is a limit for the 'height' of the parabola: It cannot exceed its 'width', otherwise the process escapes into negative values without bounds, i.e. the iteration simply breaks down. There-

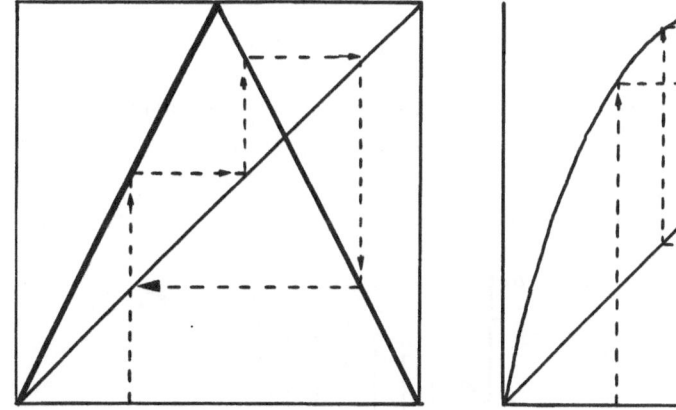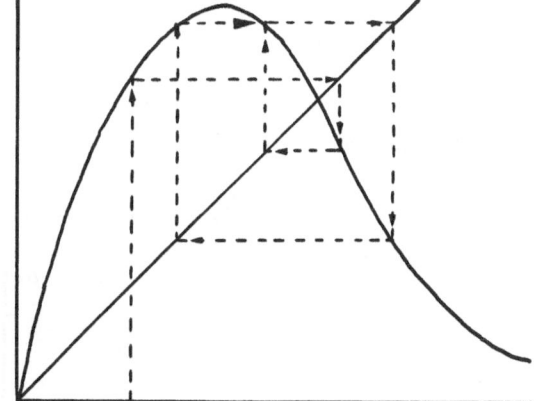

Fig. 3.5: Any convex function can be used to define an iterated map, which possesses periodic solutions. Dashed lines indicate pathways which terminate in a cyclic motion.

76

fore, we are not free in choosing the parameter r which is bound to values $0 < r \leq 4$ ($y(0.5)=r/4 \longrightarrow r_{max}=4$), and a second limit is given when the parabola is so shallow that it does not intersect with the line $y_{i+1}=y_i$. Some possible alternative models are (OSTER & GUGGENHEIMER, 1976):

$$y_{i+1} = y_i \exp(r(1-ay_i))$$

and

$$y_{i+1} = y_i/(1+\exp(-b(1-ay_i))).$$

For $r \ll 1$ the last equation can be approximated by equation (3.3) (see OSTER & GUGGENHEIMER, 1976). In general, the logistic equation provides a 'prototype' of chaotic behavior of iterated maps, which easily can be analyzed, and, therefore, it is the most celebrated example.

3.1.2   The Numerical Approximation of a Partial Differential Equation

In geology the partial differential equation $u_t = u_{xx}$ plays some role as a 'transport equation'. It describes e.g. the flow in porous media and the compaction of sediments (TERZAGHI, 1943; DESAI & CHRISTIAN, 1977). A straight forward approximation by differences leads to the explicit scheme

$$(U_{x,t+1}-U_{x,t})/\Delta t = (U_{x-1,t} - 2U_{x+1,t} + U_{x+1,t})/\Delta x^2. \qquad (3.6)$$

This equation can be rewritten as an iterated map

$$U_{x,t+1} = (1-2(\Delta t/\Delta x^2)U_{x,t} + (U_{x-1,t} + U_{x+1,t})\Delta t/\Delta x^2. \qquad (3.7)$$

It is well known from numerical mathematics (MARSAL, 1976) that the stability of this approximation requires that

$$1 - 2(\Delta t/\Delta x^2) \geq 0. \qquad (3.8)$$

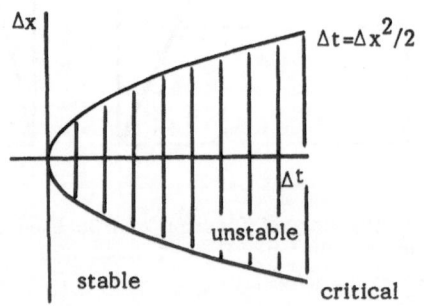

Fig. 3.6: Stability region of equation (3.8).

The first right hand term of equation (3.7) needs to be positive, and this determines the stability condition. If the left hand side of equation (3.8) is set to zero, the equation describes a parabola (Fig. 3.6) in the control space $(\Delta t, \Delta x)$, and the interesting point is what happens in this case with equation (3.7). If the control parameter (3.8) is zero, the local solution of the map (3.7) does not further depend on the $U_{x,t}$ values, and the grid separates into two disconnected substructures (Fig. 3.7). The solution then does

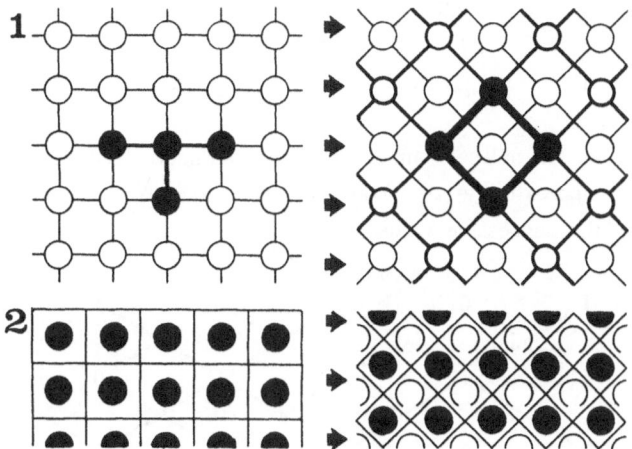

Fig. 3.7: Two representations of the grid bifurcation for the explicit difference scheme of the transport equation $u_t = u_{xx}$. The bifurcation occurs if the control parameter is zero $(1 - 2(\Delta t/\Delta x^2) = 0)$.

not further describe the original 'transport equation', but two independent solutions of this type arise. In consequence, the solution depends strongly on the initial data configuration, a fact which is sometimes not recognized (MARSAL, 1976). To see how the solution depends on the initial conditions, the degenerated version of equation (3.7) can be written as

$$U_{x,t+1} = (U_{x-1,t} + U_{x+1,t})/2 . \qquad (3.9)$$

Now, we take the example of a substance spreading from a source of constant intensity into an empty medium, and we compute this by use of equation (3.9):

t = 0	1	0	0	0	0	0	0
t = 1	1	0.5	0	0	0	0	0
t = 2	1	0.5	0.25	0	0	0	0
t = 3	1	0.625	0.25	0.125	0	0	0

The numerical approximation looks rather well, and, as can be shown (MARSAL, 1976), it really approximates the differential equation. Next, we take the same boundary conditions but different initial conditions:

t = 0	1	0	1	0	1	0	...
t = 1	1	1	0	1	0	1	...
t = 2	1	0.5	1	0	1	0	...
t = 3	1	1	0.25	1	0	1	...
t = 4	1	0.625	1	0.125	1	0	...

This time, the solution is neither numerically nor physically reasonable, we get fluctuations which cannot approximate the transport equation. What happens, becomes clear if one sketches how the successive values are connected:

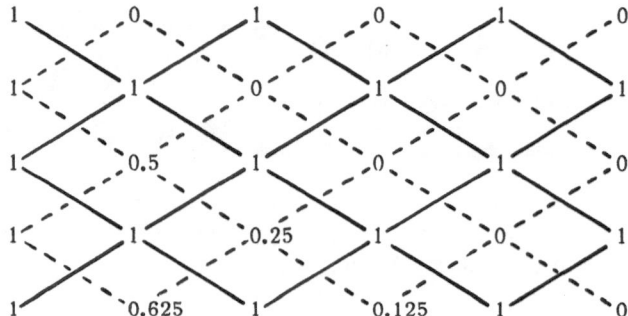

Clearly, the bifurcation of the grid into two independent subsets causes two independent solutions -- one is constant because the initial conditions are constant on this subset (the diagonal series of ones), the other one resembles the solution of the first example, i.e. a spreading process from the left boundary into an empty medium. It is not hard to see that the stable solution of the first example is not really stable, as sometimes is assumed in texts on numerical methods (MARSAL, 1976), but that it also consists of two independent solutions, which under the special conditions become identical.

In case the critical parameter (3.8) takes values less than zero, the result fluctuates and assumes negative values on one of the bifurcation grids. In this case, the grids are again connected, but the negative control parameter causes alternating signs of the $U_{x,t}$ values so that one, in principle, has still two different solutions of the discussed type. In addition, we observe a close relationship to the discussion of regular and centered grids in sections 2.4.4-5. Indeed, as the critical value of the parameter (3.8) is approached, the stable regular grid (Fig. 3.6-1) evolves into two grids which resemble the centered grids of the previous discussion. These grids are disconnected and provide two independent solutions. The stability problem, therefore, has a strong topological component. Another analogy provides the discussion of linear systems in section 2.2.1, where we observed a similar parabolic stability boundary. Of course, the partial differential equation $u_t + u_{xx} = 0$ can be approximated by a set of differential equations

$$
\begin{aligned}
y'_1 &= a_{11}y_1 + a_{12}y_2 + \cdots + a_{1n}y_n \\
y'_2 &= a_{21}y_1 + a_{22}y_2 + \cdots + a_{2n}y_n \\
&\text{etc.,}
\end{aligned}
\tag{3.10}
$$

which provide a discontinuous spatial but continuous temporal approximation.

To generalize this result, we can briefly analyze the difference approximation of first derivatives. There are three approximations in use (e.g. DESAI & CHRISTIAN, 1977), the

forward difference	$(u_{i+1,\ j} - u_{i,\ j})\ /\Delta x + O(\Delta x)$
backward difference	$(u_{i,\ j} - u_{i-1,\ j})/\Delta x + O(\Delta x)$
central difference	$(u_{i+1,\ j} - u_{i-1,\ j})/(2\Delta x) + O((\Delta x)^2).$

Under numerical aspects the central difference should be the best one to approximate a first partial derivative because its discretization error is only of order $(\Delta x)^2$. But nearly all approximations using the central difference are instable (MARSAL, 1976). The previous discussion has shown that this is not a numerical problem but a topological one. The central difference causes a grid bifurcation as in the previous example, i.e. one computes two independent solutions, and, therefore, the approximation can only be used for very special initial conditions. Actually, the problems, which arise here, are very close to those discussed in the last chapter, especially the surface approximations from scattered data.

3.1.3   Infinite Series of Caustics

In refraction seismology one is sometimes interested in the so-called 'higher arrivals' and in the caustic formed by the rays. The caustic is the envelope of rays, which, in its totality, can be written as

$$
F(x,y,p) = 0 .
\tag{3.11}
$$

The equation of the envelope of this family of rays is obtained by eliminating the ray parameter p from equation (3.11) and from its partial derivative

$$\frac{\partial F(x,y,p)}{\partial p} = 0 \qquad (3.12)$$

(e.g. BEN-MENAHEM & SINGH, 1981). For a general overview one can choose special conditions, i.e. the situation where the source is located at the reflector. For a medium with linear velocity increase, the rays are circles (e.g. OFFICER, 1974). Because the aim here is only to demonstrate the use of iterated maps, a still more simple model will be used, parabolic rays. If the curvature of the rays under consideration is large, the parabolic rays approximate the circular ones to some extent. If all parabolic rays have a common source point, equation (3.11) becomes

$$y - (x^2 - bx) = 0. \qquad (3.13)$$

A specific ray with ray parameter b passes through the points $x = 0$ and $x = b$ if $y = 0$. At $x = b$ the ray is reflected, and because the source is located at the reflector, it reaches the reflector a second time at $x = 2b$. This gives the general iterated map for the reflection points

$$x_i = x_{i-1} + b \qquad (3.14)$$

or

$$x_i = x_0 + ib.$$

The ray itself is defined in the interval (0,b) and maps iteratively into the intervals (b,2b), (2b,3b), ... Or if a certain interval is given, the equation for the rays (3.13) takes the form (by use of equation (3.14))

$$y - ((x_i - ib)^2 - b(x_i - ib)) = 0. \qquad (3.15)$$

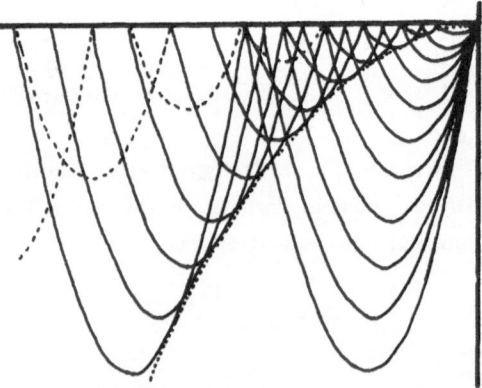

Fig. 3.8: Caustic(s) of a simple parabolic ray system with reflection at the surface. The source is located at the reflector.

By use of equations (3.11) and (3.12) one finds the equation of the caustics by eliminating the ray parameter b:

$$y - x_i{}^2(1-(1+2i)^2/(4i(i+1))) = 0$$

or   (3.16)

$$y + x_i/(4i(1+i)) = 0.$$

Thus, the caustics are an infinite series of parabola with increasing curvature, which all pass through the point (0,0). Fig. 3.8 gives the first iteration as an example. More complicated ray systems like circular rays or arbitrary positions of the source can be treated in the same way.

Of some interest is the case of a source located below the reflector. This situation causes a bifurcation, which can be qualitatively described in the following way. The source is now located at depth 'a' and the reflector, as before, at depth zero. Then the ray equation (3.16) becomes

$$(y+a) - (x^2-bx) = 0,$$ (3.17)

i.e. all rays are passing through the source point (-a,0). Now, the rays are reflected at y = 0, and their horizontal position is then

$$-x^2 + bx +a = 0.$$ (3.18)

To find the reflection points, one has to solve the quadratic equation (3.18), and, in general, this will give two reflection points because the ray propagates into the positive and into the negative x-direction from the source point (o,-a). Because of symmetry

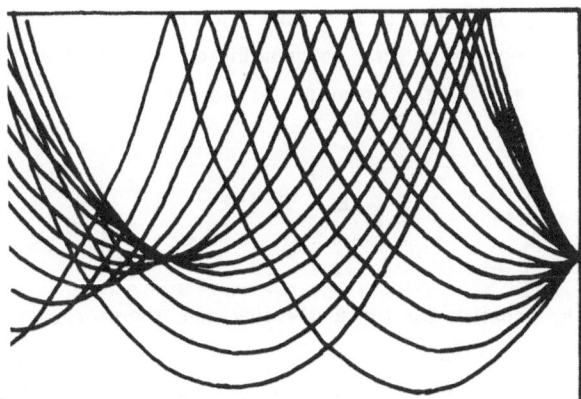

Fig. 3.9: Caustic of parabolic rays. The source is located below the reflector.

reasons, we have to consider the absolute values of the reflection point; both solutions propagate into the positive and into the negative direction -- we have to fold the solution space. Only in this case, the whole semiinfinite solution space is covered with rays. The consequence is that the ray parameter 'b' is not further uniquely defined; it describes two rays, and both ray systems are capable to produce iteratively caustics. The two caustics are connected at a cusp point, i.e. one finds cuspoid caustics (POSTON & STEWART, 1978; Fig. 3.9). The bifurcation of the solution arises also with other ray systems, such as circular rays. It is not a property of a special ray system, but it only depends on the position of the source; it is a structurally stable topological property of ray systems.

3.2   CHI^2-TESTING OF DIRECTIONAL DATA

Within the analysis of orientation data one has to prove whether the observed distribution has pronounced extrema. The usual way is to test the contrary, whether the data differ significantly from a uniform distribution. The propagated method in text-books (e.g. MARSAL, 1979) is the Chi^2-test. BALLENTYNE & CORNISH (1979) observed that the Chi^2-value obtained from such a test depends on the arbitrary selection of the offset point for the sector system, in which the directional data are grouped. By an extensive numerical analysis they showed that the Chi^2-values fluctuate as the sector pattern is rotated over the data. In their analysis, the Chi^2-values passed thereby several times the significance level. Therefore they concluded:

> *"The hitherto widespread use of the test in this way cannot, therefore, be considered a valid mode of analysis, and the results of previous studies employing this methodolgy must be treated with extreme caution."*

While the numerical analysis of BALLENTYNE & CORNISH (1979) shows that this is a problematic test, it does not elucidate why the uniform distribution as a zero-hypothesis behaves in such an unpredicted way. Ballentyne & Cornish noticed that the expected frequency E is a constant for all sectors under the special condition of a uniform distribution:

$$E = N \: / \: k; \qquad \qquad \text{N: number of data} \qquad (3.19)$$
$$\text{k: number of sectors.}$$

The equation for the Chi^2-value can, therefore, be written as

$$\chi^2 = \frac{1}{E} \sum_{i=1}^{k} ((O_i - E)^2 \quad = \sum_{i=1}^{k} (O_i - E_i)^2 / E_i \qquad (3.20)$$

$$\text{if } E_i = \text{constant;}$$
$$O_i: \text{number of observed data in the sector i.}$$

But this simplified version contains still a constant value inside the sum. Further evaluation yields

$$\chi^2 = \frac{1}{E} \sum_i O_i^2 - 2 \sum_i O_i + kE. \tag{3.21}$$

The indices of the sums are the same as in equation (3.20), they will not be repeated in the following equations. By use of equation (3.19) one has the relations

$$O_i = N \quad \text{and} \quad kE = N ,$$

and this gives finally

$$\chi^2 = \frac{N}{k} \sum O_i^2 - N. \tag{3.22}$$

Thus the obtained Chi^2-value depends only on the sum of squares of the observed values in the sector system. Because N and k are constants for a certain data set and a given sector pattern, one can define a modified test statistic , which simplifies the further analysis:

$$\frac{k}{N} \chi^2 + 1 = \sum O_i^2 . \tag{3.23}$$

The important point is that the test statistic does not really depend on the expectation values E_i; the only remaining variables are the O_i's, and the O_i's can be altered by a dislocation of the sector system or by another spacing of the sectors. The same situation arises in normal histograms and two-dimensional data if they are tested against the uniform distribution. The derived folmulae also hold in these cases. To study the behavior of the test statistic it will be sufficient to alter the offset point of the sector system.

Rotation of the sector pattern causes jumps of data points from one sector into the neighboring one when the sector boundary passes a data point. Such a jump modifies the sum $\sum O_i^2$ locally:

$$(O_i-1)^2 + (O_i+1)^2 = O_i^2 + O_{i+1}^2 + 2(O_{i+1}-O_i) + 2. \tag{3.24}$$

For an arbitrary number of jumps between two classes one finds

$$(O_1-J)^2 + (O_2+J)^2 = O_1^2 + O_2^2 + 2(O_2-O_1) + 2J^2 \tag{3.25}$$

$$J: \text{ number of jumps.}$$

Therefore, the test statistics is changed by a value

$$2(J(O_2-O_1) + J^2).\tag{3.26}$$

The general equation for a dislocation of the sector pattern is found by summing up all local changes, and the modified test statistic (3.21) can be written as

$$\frac{k}{N}\chi^2 + 1 = \sum O_i^2 + 2\sum O_i(J_i-J_{i+1}) + 2\sum J_i^2 - 2\sum J_iJ_{i+1},\tag{3.27}$$

where the sector $k + 1$ is identical with the sector 1. The equation shows that any jump of a data point from one sector into another, under rotation of the sector pattern, will alter the test value, as was numerically found by Ballentyne & Cornish. But equation (3.27) allows a more detailed analysis. To have no change of $\sum O_i^2$ under a rotation of the sectors requires

$$0 = \sum O_i(J_i-J_{i+1}) + \sum J_i^2 - \sum J_iJ_{i+1}$$

or alternatively $\tag{3.28}$

$$0 = \sum J_i(O_{i+1} - O_i) + \sum J_i^2 - \sum J_iJ_{i+1}.$$

These conditions can only be satisfied if $J_1 = J_2 = ... = J_k$, or if $J_i = 0$ for all i. In all other cases, the original sum is altered, and one gets different test values. Now, the condition to have an equal number of jumps for all sectors is mainly a geometrical problem.

The constraints given by equation (3.28) require that the data are equally spaced on the circle and that they have unique frequency. The unique frequency within every sector is what one suspects to be proved by the test method, but now spacing appears as a new parameter, which affects the test value. In addition, on a sufficiently fine scale the unique distribution does not play any role furthermore. The data are measured on a scale of real numbers, and, therefore, any local cluster is due to the roughness of the measurement; as the scale becomes finer and finer, the cluster will be divided into single points. Therefore, on a sufficiently fine scale the data are single points on the circle (Fig. 3.10). The only remaining variable, then, is the spacing of the data points. One can interpret any theoretical distribution in the way that the density over

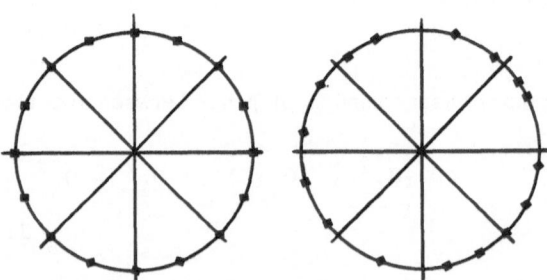

Fig. 3.10: Equally spaced and uniform distributed points on a circle.

a certain interval gives a measurement of the (infinitesimal) spacing of points on the real line. In the case of the uniform distribution, the density describes an (infinitesimal) uniform spacing. But what we expect, are not equally spaced data. Our opinion is that the data result from a random process, which selects the data with equal probability from a certain interval of the real numbers. Therefore, one cannot expect to find equally spaced data in a finite size sample -- one has, of course, the combinational problem to arrange N data on M points on the real line where M is given by the roughness of the measurement.

Now, one may ask how strong the alterations of the test value for different distribution patterns are. The most simple system are two sectors, and 100 data give a likely sample size. In a first case, we may have nearly equal spacing, then every sector contains approximately 50 data points. The rotation of the sector system causes only jumps of a few data points at once, say, in the order 1 to 10. We find that $\sum_i 0_i$ takes values like (equation (3.24)):

$$
\begin{aligned}
50^2 + 50^2 &= 5000 \\
49^2 + 51^2 &= 5002 \\
&\cdot \\
45^2 + 55^2 &= 5050 \\
&\cdot \\
40^2 + 60^2 &= 5200.
\end{aligned}
$$

The change of the test value is not very impressive. Changes of this magnitude are found if random numbers are • used to provide a numerical test. Now, assume in the second case that the distribution has a single well pronounced maximum so that it is possible to locate all data in one of the two sectors. On the other hand, there exists a rotation of the sector system, which divides the data into two sets with nearly equal frequency; therefore, under the rotation one will find sums in the range

$$
\begin{aligned}
0^2 + 100^2 &= 10\ 000 \\
&\cdot \\
50^2 + 50^2 &= 5\ 000,
\end{aligned}
$$

and the extrema diverge quite clearly. For this data configuration the behavior of the system becomes rather chaotic with respect to an arbitrary location of the sector system. In addition, the possible outcomes diverge rapidly with increasing sample size. Thus, we have the striking situation that, in contradiction to the general statistical opinion, an increase of sample size does not stabilize the result, but that the contrary is true:

The uncertainty about the result of the test increases as the deviation of the sample from the uniform distribution approaches certainty.

These observations allow a final discussion of the causes for the instability of the test. Take a two-sector sample from the normal distribution over the circle (e.g. MARDIA, 1972), then there are two extreme cases: In the first, one sector contains nearly all data, in the second, both sectors contain an equal number of data. This is again the situation considered above. But now, with the normal distribution, the expectation values are neither independent of the data structure -- they depend on the mean and variance of the data -- nor are they independent of the choice of the offset point of the sector system. As the sector rotates, the observed frequency within a sector is altered in the same way as the expectation values for this sector and vice versa. To use a term from synergetics (HAKEN, 1977), both values are 'slaved' by the position of the sector system. In the case of a uniform distribution, the expectation values break out of this 'slaving', and this allows the system to fluctuate free as described above. The 'revolt of the slaved parameters' becomes especially strong when the observed frequencies are strongly dependent on the position of the sector system, i.e. when the distribution has a well pronounced maximum.

3.3 PROBLEMS WITH SAMPLING STRATEGIES IN SEDIMENTOLOGY

The analysis of pseudo-time series (stratigraphic thickness against some variable) by means of classical methods like polynomial curve fitting, moving averages, cross correlation etc. is well known (FOX, 1975; SCHWARZACHER, 1974). Besides these methods, random models have been used, and here especially the concept of transitional probabilities and of Markov chains (KRUMBEIN, 1975). In sedimentology and stratigraphy the general problem with these techniques is that the 'time series' or the 'sequence of signals' is usually too short because the profiles are of limited length. Other problems result from special, propagated sampling techniques like equal distance sampling. These problems are of special interest in the present context because they cause problems of convergence, and they are capable to generate artificial patterns: They are, in some way, the inverse problem of iterated maps. It will turn out that the problems are again geometrical ones, and that it is not advisable to replace geological (morphological) reasoning by a sampling formalism.

3.3.1 Markov Chains in Sedimentology

There arise special problems if transitional probabilities are used to study periodicity patterns of profiles. These problems are mainly of a classificatory nature; they resemble closely the problem to define the geometry of a bed or facies unit. Furthermore, they are related to the process of sedimentation and from this to the type of 'signals' --

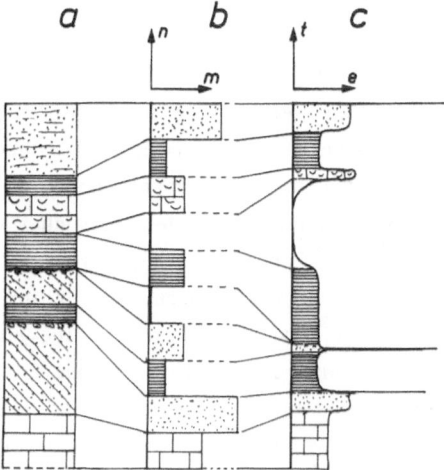

Fig. 3.11: Graphic representation of different scales for definition of probabilities on profiles. a) The profile in classical representation, b) the probability to find a certain lithology measured in terms of bed thickness, c) the probability of a lithology to be deposited during a time interval.

whether the sedimentation process consists of distinct events or reflects a continuously changing environment.

Transitional probabilities are just one possible definition of a large set of probability measurements to be defined on profiles. Some other types of probability measurements on profiles may be briefly reviewed as a base for the later discussion of transitional probabilities. From a sedimentological viewpoint, we have the total probability of a certain facies type, the likelihood to take a sample from the profile and to find a sandstone, carbonates, a claystone etc. In order to define the probability measurement, the profile has to be classified into distinct facies units or beds, and the probabilities result from bed-thicknesses. Still the same probability structure but with other probability values results if the depth scale of the profile is transformed into a time scale. So, any smooth deformation of the scale will give new probability values, but it will not disturb the special probability algebra (Fig. 3.11). Another type of probabilities with somewhat different algebraic rules (RENYI, 1977; FISZ, 1976) results if two objects are analyzed simultaneously. In this case, one works with conditional probabilities, e.g. the probability to find a certain fossil in a specified lithology. If one can define some ordering relation like before and after, then a series of discrete signals can be analyzed in terms of the conditional probabilities: to find a certain signal before or after another one. If the positional relationship is reduced to 'just before' or 'just after', then the conditional probabilities become the classical transitional probabilities of a Markov chain. The structure of the conditional probability space depends not only on the scale used for the measurements but also on the definition of the relationship between the two

88

sets of objects.

In sedimentological practice the main problem is to define distinct signals, i.e. distinct sedimentological units. In a very narrow sense, this is only possible if the sedi- mentation process is not continuous. But long series of sedimentation by events usually are restricted to turbidites, in which the sedimentological structure is very homogene- ous -- commonly without transitions between a larger number of lithologies. In all other cases, the transitions between beds or facies units are more or less continuous. But, as sharp boundaries between the 'signals' disappear, the definition of a probability space becomes more and more subjective, and this contradicts the aim of an objective statisti- cal analysis. For this reason, a modified sampling technique is frequently used to establish the empirical transition probabilities of a 'sedimentological' Markov chain (MIALL, 1973). Instead of discrete signals, which have to be defined subjectively, one takes small samples in some regular distance. The reason is that the samples can be more easily classified. It seems worthwhile to analyze how the different sampling methods may influence the probability structure.

A) Discrete Signals

First some additional features of the classical approach of Markov chains may be discussed with respect to the sedimentological questions. A sedimentary unit character- ized by its lithological, sedimentological, paleontological etc. content needs to be defined as a signal, which is an event either a priori -- separated from the events 'just below' and 'just above' by distinct boundaries -- or which can become a 'distinct event' by a useful definition of its boundaries. Because the event is defined by its structure, the transition matrix is independent of bed thickness and profile depth. The transition matrix is of the form

$$\begin{pmatrix} \cdot \cdot \cdot & \cdot & \cdot \cdot \cdot \\ \cdot \cdot \cdot & p_{ij} & \cdot \cdot \cdot \\ \cdot \cdot \cdot & \cdot & \cdot \cdot \cdot \end{pmatrix} = \begin{pmatrix} \cdot \cdot \cdot & \cdot & \cdot \cdot \cdot \\ \cdot \cdot \cdot & n_{ij}/ N_j & \cdot \cdot \cdot \\ \cdot \cdot \cdot & \cdot & \cdot \cdot \cdot \end{pmatrix} \qquad (3.29)$$

n_{ij}: number of transitions from signal S_j to S_i;
N_j : total number of occurrences of the signal S_j.

Repetitions of identical units such as a sequence sandstone --- sandstone are usual events in classical Markov chains. In sedimentology they are only possible if there exists some boundary which can be recognized, a feature which usually is related to banking. Now, the boundary between two beds, no matter how small, means some change in sedimentation, e.g. a short interval of lowered sedimentation rate that caused the physical boundary. Thus, if we do not identify the boundary between beds as a separate

89

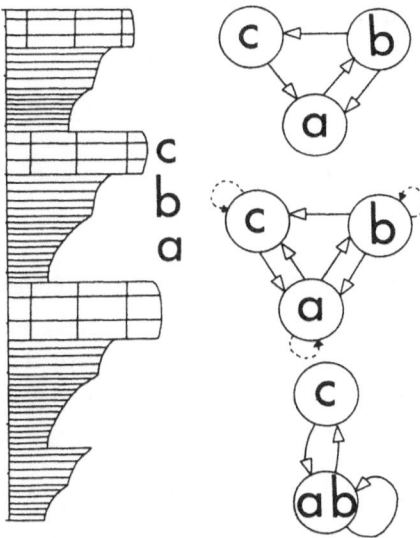

Fig. 3.12: A sequence of three sedimentary units a, b, c and different possible transition graphs. Above: transitions if the sedimentary units are defined as 'distinct events', middle: one possible transition graph for equal interval sampling; the number of singular loops depends on the spacing of the samples. Below: occurrence of repetitions, i.e. singular loops, due to lumping of lithologies; 'a' and 'b' are not distinguished.

event, this may be due to a high threshold that has disturbed the record. This view can be formalized in terms of a map from the transition matrix of degree n onto a transition matrix of degree n - k, e.g.

$$\begin{pmatrix} 0 & p_{12} & p_{13} \\ p_{21} & 0 & p_{23} \\ p_{31} & p_{32} & 0 \end{pmatrix} \xrightarrow{\quad S^3 \dashrightarrow S^2 \quad} \begin{pmatrix} p_{11} & p_{12} \\ p_{21} & p_{22} \end{pmatrix}. \tag{3.30}$$

A more instructive representation of this map can be given by a transition graph (Fig. 3.12) which shows how singular loops, like a sandstone --- sandstone sequence, arise due to ignorance or to lumping of intermediate signals (e.g. sandy shales). As will be discussed below, the same structure with singular loops arises if the samples are taken in equal intervals. Within sedimentological problems, the occurrence of singular loops usually indicates that a transition state (such as non-deposition) is missing.

B) Equal Interval Sampling

If we now go on to the second sampling method, the sampling in discrete intervals, we have to prove whether it really terminates into a Markov chain. MIALL (1973) writes that this

> *"method can give rise to a much more accurate measure of the relative frequencies of the lithotypes present, but at the expense of accuracy in measuring step-by-step depositional changes".*

This remark shows that one has to take care that the probabilities are well defined, i.e. that one does not produce a mixture out of conditional and total probabilities (see above). Miall further emphasizes a sampling interval slightly less than the average bed thickness. This additional recommendation will be studied in detail in the next section. In order to analyze the interval-sample strategy one may assume to have a series of well defined events with transitional probabilities

$$
\begin{pmatrix}
0 & \cdots & \cdots & \cdots \\
\cdots & 0 & p_{ji} & \cdots \\
\cdots & p_{ij} & 0 & \cdots \\
\cdots & \cdots & \cdots & 0
\end{pmatrix}
=
\begin{pmatrix}
0 & \cdots & \cdots & \cdots \\
\cdots & \cdots & n_{ji}/N_j & \cdots \\
\cdots & n_{ij}/N_i & 0 & \cdots \\
\cdots & \cdots & \cdots & 0
\end{pmatrix}
\tag{3.31}
$$

The samples are taken in equal intervals from the identical universe. As long as the sampling distance is larger than the average bed thickness, one will miss quite a lot of transitions. As the sampling distance becomes less than the smallest bed, the counts of transitions stabilize, but the sampling distance can be further reduced, in the extreme case down to an infinitesimal small size. From the point, where the transition counts (not yet probabilities) become stable, one will find new transitions only along the diagonal of the counting matrix. The counts along the diagonal will increase with decreasing sampling distance until the values equal the total thickness of every lithotype. The sum of the diagonal elements, therefore, gives the length of the profile:

With respect to the content of the diagonal elements, therefore, every interval-sampling process generates nothing but a measurement of the total thickness of the different lithotypes within a profile -- measured as meter-intervals, cm-intervals or, generally, in units of the sampling distance.

The elements of the diagonal line, divided by their sum, provide the total probability to find a lithotype within the profile. If one computes transition probabilities from this counting matrix, then the probability matrix reads

$$\begin{pmatrix} \cdots & \cdots & \cdots & \cdots \\ \cdots & m_{ii}/(m_{ii}+N_i) & \cdots & \cdots \\ \cdots & \cdots & n_{ji}/(m_{jj}+N_j) & \cdots \\ \cdots & \cdots & \cdots & \cdots \end{pmatrix} \qquad (3.32)$$

m_{ii}: thickness of the lithotype i;

n_{ji}: counts of transitions from lithotype j to i;

N_j: total number of transitions from lithotype j to another state.

This probability matrix has a suspicious structure. First, the probabilities are not dimensionless numbers because the n_{ij}'s are counts while the m_{ii}'s have the dimension of a length measurement, and this is independent of the roughness of the intervals -- but, per definition, probabilities are dimensionless numbers. On the other hand, if we take a small sampling distance on a profile, where the bed thickness is not small relative to profile length, the diagonal elements approach one, and the transition graph (Fig. 3.12) is dominated by singular loops.

The suspicious structure of this probability matrix becomes clear if one separates the counting matrix into its independent parts

$$(3.33)$$

$$\begin{bmatrix} \cdots & \cdots & \cdots & \cdots \\ \cdots & n_{ii} & n_{ji} & \cdots \\ \cdots & \cdots & \cdots & \cdots \end{bmatrix} = \begin{bmatrix} 0 & \cdots & t_{ij} & \cdots \\ t_{ji} & 0 & \cdots & \cdots \\ \cdots & \cdots & 0 & \cdots \end{bmatrix} + \begin{bmatrix} \cdots & 0 & 0 & 0 \\ 0 & m_{ii} & 0 & 0 \\ 0 & 0 & \cdots & 0 \end{bmatrix}$$

The t_{ij} are the real transitions between lithotypes, the m_{ij} measure the thickness of the individual lithotypes.

From the matrices on the right side one finds two independent probability systems -- the transitional probabilities from the t_{ij} and the probability or relative frequency of a lithotype $m_{ii}/\sum_i m_{ii}$. It seems, therefore, rather dangerous to transfer the geological or sedimentological decision 'what can be defined as a discrete lithological signal' to a pseudo-objective sampling strategy. What is the meaning of the 'probability matrix' n_{ij} when the thickness of the beds varies strongly? In the case that the beds have nearly the same thickness, the additional rule that the sampling distance should be taken near the average bed thickness secures that the estimated matrix approximates the transition matrix to some extent. In the case of highly variable bed thickness, the matrix n_{ij} will not make much sense. At least, if one can compute a mean bed thickness, one has some idea what a bed looks like in the profile under study, and why should one then go this doubtful way?

3.3.2   Artificial Pattern Formation in Stratigraphic
Pseudo-Time Series

It is known for a long time that several astronomic cycles in the order of 20 000 to 400 000 years may cause climatic changes, and there has been a large number of attempts to relate geological phenomena to these astronomic cycles, e.g. there are several good arguments in the case of marl-limestone rhythms, coming from the proposed climatic changes (EINSELE & SEILACHER, eds. 1982). However, it is very hard to give a statistical proof of the correlation between bedding phenomena and astronomic cycles. To do this would require to have true time series of carbonate production and of clay influx under controlled conditions. But already the relation between time and sediment accumulation is not known down to sufficiently small intervals in any profile. Therefore, this problem is usually ignored and a fairly constant rate of sedimentation assumed (SCHWARZACHER & FISCHER, 1982). An additional handicap is that most computer procedures for time--series analysis require equal interval samples. SCHWARZACHER & FISCHER (1982) recommend equal interval samples at a distance which is not smaller than the thinnest beds encountered with some frequency. Surely, the bed is the (possible) unit for an analysis of cycles. It is the smallest visible fluctuation within the profile. If one has enough in-formation about the cycles, one will get a good picture of the periodic process if one chooses the sampling distance exactly as half or as one wavelength. If, in addition, the sampling points are located at the extrema of the periods, one gets a very simple linear-ized approximation. For the stratigraphic problem this would require to locate the sampling points at the extrema of a limestone-shale sequence at the boundaries and the centers of beds, in order to describe the fluctuations of the carbonate content properly. But, if the data have to be analyzed with a computer algorithm, equal distant sampling points are necessary while bed thickness usually fluctuates. So, the computer obtrudes a certain strategy, whether we like it or not. The question is what we have to do in order to get a physically or sedimentologically reasonable result and not only a meaningless product of the computer. To elucidate some problems, two different aspects will be discussed, the sampling of periodic functions and the analysis of bed thicknesses. In the first case, it will turn out that the propagated sampling distances can cause artificial pattern forma-tion, in the second example, we will see that a not properly defined statistical hypothesis causes interpretational problems.

A) Sampling of Periodic Functions

SCHWARZACHER & FISCHER (1982) recommend equal interval samples at a distance which is not smaller than the thinnest beds encountered with some frequency. If we assume the bed to represent approximately one wavelength of the smallest periods, we can study the influence of the sampling distance for an idealized periodic process. A simple cosine

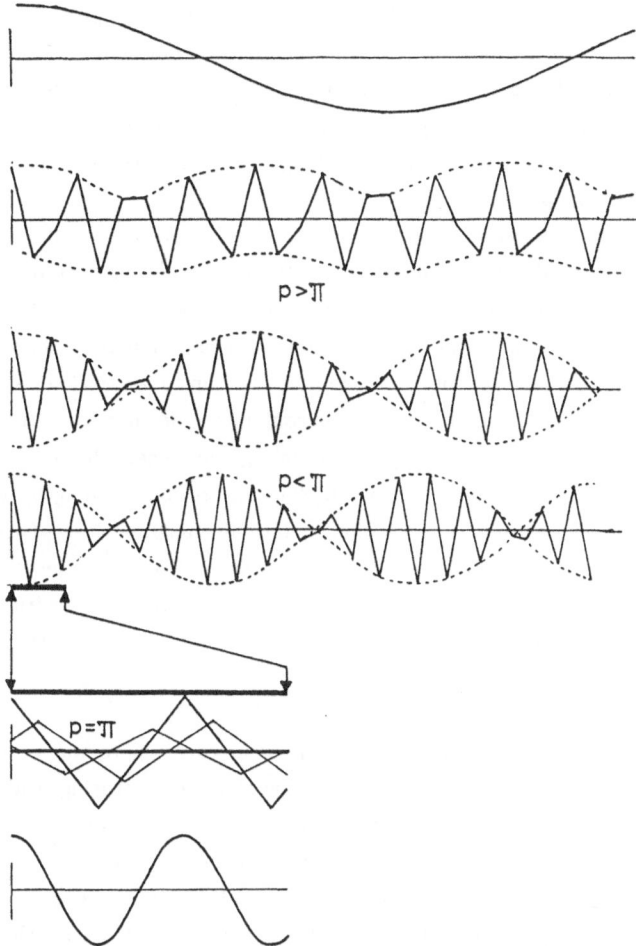

Fig. 3.13: Artificial pattern formation due to interval sampling (interval width near π) on a cosine signal. The cosine signal and the sampling patterns, which result from intervals of exactly the width π (but with different starting point), are drawn enlarged.

function provides a model for, e.g., the fluctuation of the carbonate content (Fig. 3.13). Sampling at distances of exactly π will be successful if the starting point of the samples does not coincide with the inflection point of the cosine function. In this special case, no periodicity will be detected -- anyway, if the sampling distance is chosen as 2π, then the periodicity will vanish for every starting point. For all other starting points, one finds amplitude fluctuations which correctly represent the wavelength, but the true amplitude is only found if the sampling points coincide with the maxima of the cosine function.

Now, it is unlikely that the sampling intervals have exactly the length π (or 2π)

even in a computer simulation. Therefore, we disturb the sampling distance slightly, i.e. we take distances $\pi + \varepsilon$ (or $2\pi + \varepsilon$). What happens, demonstrates Fig. 3.13. The original cosine function is modulated, and new periodic patterns of higher order occur which only depend on the disturbance parameter ε. In the extreme case, a new simple periodic curve appears with several times the wavelength of the original signal. The error term ε causes the sampling points to 'move' along the periodic function, and depending on the error parameter this shifting process adds an amplitude modulation to the cosine function. Already for small values of the error term we get an infinite number of possible higher periods and of pure chaotic behavior -- this depends on the quotient $n\pi/(\pi + \varepsilon)$; if there exists a 'n' such that the quotient is a rational number, then we have a period of $n\pi$; otherwise, we have chaotic behavior. Now, in a numerical sense, the sampling distance near π is much too large to approximate the periodic function; an analysis by equally spaced samples would require sampling distances of only a fraction of π. As the sampling distance becomes small in relation to phase length, we can safely use equal distance sampling as well as the mathematical 'machineries' of time-series analysis and control theory. In geology, however, we have to consider the sampling scale, and this relates the problem to iterated maps, where similar problems with distances occur -- like in the example of the finite logistic equation. The relation to iterated maps can be illustrated in more detail.

The cosine function is a well defined map of a rotating radius vector of unit length onto a Cartesian coordinate system and vice versa. The rotating radius vector can be identified with a mass point, which moves with constant velocity on the circle. Equal distance sampling, then, is equivalent to a uniform motion of the mass point inside the circle with regular elastic collisions with the circular wall. The points of collision are the sample points. This is a classical example of chaotic motion (GRENANDER, 1978): If the first angle of collision is denoted by θ (measured from the radius vector of the circle), then the arc length between successive impacts is $\pi - 2\theta$ (from planimetry). The

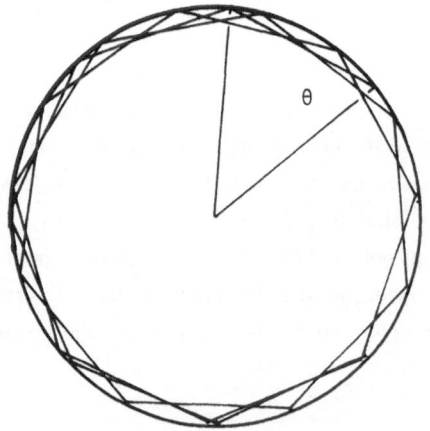
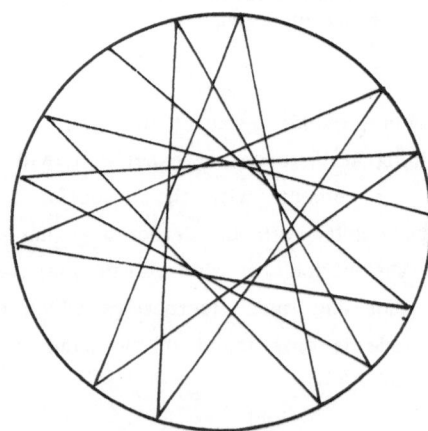

Fig. 3.14: Elastic motions in a circle.

resulting pattern depends on the ratio $2\pi/(\pi-2\theta)$. If 2θ and π are incommensurable, the system tends to long term chaotic behavior (Fig. 3.14); however, the system depends also on the magnitude of θ. If θ is large, we have a high frequency of collisions, and the pathway of the mass point approximates the circle. If θ is small, a quite different pattern appears, like in Fig. 3.11b, where the pathway consists of quasi triangles, which slowly rotate. The lines of motion envelop again a circle, and in terms of the approximation problem the relation of the radii of the outer and inner circle provides a measurement for the quality of approximation (the outer circle is the function which should be recorded).

B) The Analysis of 'Bed Thickness' by Equal Distance Samples

In the previous example we had two variables, profile length and a second independent variable, for instance carbonate content. If rhythms are studied, usually the only variable is bed thickness -- the sum of this variable being profile length. Therefore, it is classically assumed that the beds have been deposited in equal time intervals, i.e. the number of beds is taken as a relative measurement of time. Then bed thickness is a measurement of the varying sedimentation rate, and these two variables form an independent frame. An alternative sedimentological model is (SCHWARZACHER & FISCHER, 1982) that the sedimentation rate was fairly constant, and that the bedding phenomenon is due to fluctuations in the lithological composition, as discussed in the last example. To analyze this model in terms of bed thickness by computer algorithms, Schwarzacher & Fischer used a special sampling technique. They positioned their equally spaced sampling points along the profile and then measured the thickness of the bed which was hidden by a sampling point (Fig. 3.15a,b). Therefore, some beds are lost while others are recorded several times. The method has the effect that it appears more simply to decide which of several possible bedding planes has to be taken as the boundary of the bed (there can be secondary features due to solution). In addition, the method allows to hope that errors will be smoothed out by the objective sampling procedure. In the next step, they related bed thickness to profile length. The resulting graph (Fig. 3.15c) can well be analyzed by methods like autocorrelation and spectra. Now, one can ask again what happens if the sampling distance becomes very small. For an 'infinitesimally' small sampling distance, the graph resembles a step function (Fig. 3.15d), and 'bed thickness' appears twice in this graph, i.e. the beds are represented as squares along the profile. It is hard to identify this representation with a useful interpretation. The only significant patterns are the steps at the boundaries of the bed. Using only these interruptions (Fig. 3.15e) one gets a series of phase modulated signals along the profile. These signals -- the bedding planes -- could be recorded directly along the profile by sedimentological reasoning, and then, of course, with higher precision than by an 'equal interval method', where the distances are chosen by a rule like 'take the thinnest beds which

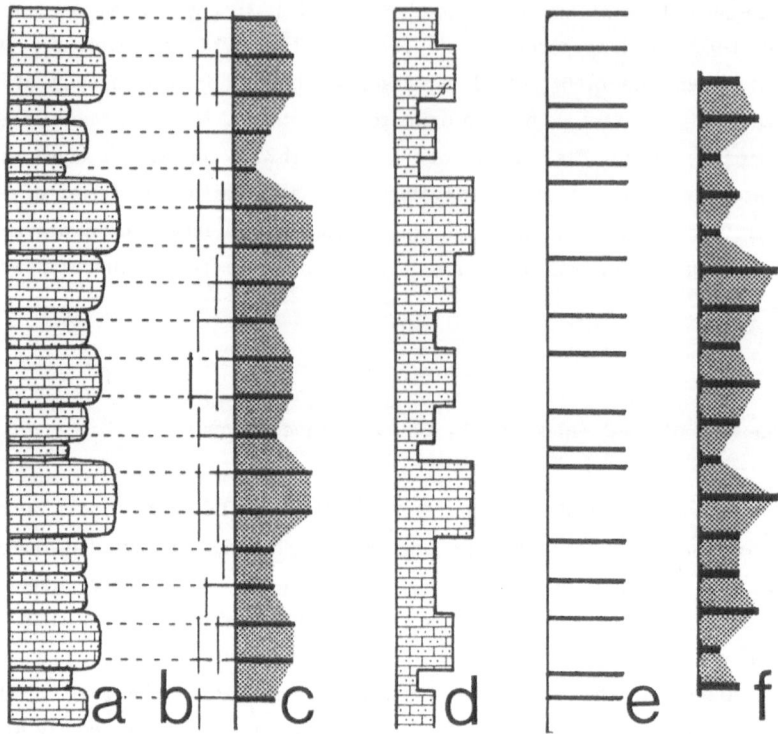

Fig. 3.15: The analysis of 'bed thickness' along a profile (a). The points, at which bed thickness is measured, are taken in equal intervals (b) and plotted against profile length (c). The transition to infinitesimally small sampling intervals gives a sequence of 'squares' (d) -- the remaining information are the interrupts between beds (e). The assumption of fairly constant sedimentation rates allows to transform the phase modulated signals (e) into amplitude modulated ones (f).

are encountered by some frequency'. Anyway, the decision is necessary, what is and what is not a bedding plane.

The only point, which is contrary to the geological method (the definition of bedding planes), is that the signals are not equally spaced, and that they, therefore, cannot be analyzed with standard computer programs for time series. On the other hand, that can be easily done if the equal distance sampling method is used -- but can this be a reason to use the less suited method, are we really 'slaved' by the computer? By a simple manipulation, the phase modulated system of bedding planes (Fig. 3.15e) can be brought into a form which allows us to analyze the data by an 'equal step' auto-correlation program. The approach, so far, was that the sedimentation rate was considered fairly constant, i.e. the distance between the bedding planes provides a measurement of the duration time of a sedimentological 'signal' -- of a bed. Therefore, one can give a plot 'number of the initial signal' or 'number of the interrupt' against the 'duration time of the signal' (Fig. 3.15f). This transforms the phase modulated signal of bedding planes into an amplitude modulated one with equal distances between the data points. But now, it turns out that the two different sedimentological models become identical. The assumption that the sedimentation rate is fairly constant cannot be distinguished from the model that the beds are deposited as events in equal time intervals because the duration time, as defined, is proportional to bed thickness. This, of course, holds only for the 'statistical approach'; by geological reasoning it is usually possible to distinguish these two cases (EINSELE & SEILACHER, eds. 1982). Thus, one comes out with the result that the connection of the depositional models with a statistical formalism can cause a worse defined statistical problem, in the sense that the statistics cannot prove which of the sedimentological models is the correct one while a successful statistical test always seems to prove the a priori sedimentological model. The problem whether an initial model is consistent with the statistical analysis is not a trivial one; in topology the analogous problem is genericity.

"To many scientists the 'genericity' problem has always been interesting. What we are looking for is the following: given a mapping $f:U \rightarrow R^m$, where U is an open set in R^n, how can we perturb f slightly to obtain a nicer and simpler mapping?" (LU, 1976).

The idea to analyze the sedimentological problem in terms of a mapping leads to the following diagram

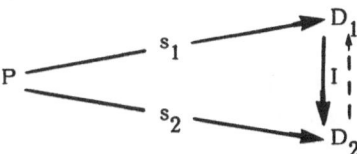

The set of observations (location and thickness of beds: P) maps under the sedimentological hypothesis s_1 (event sedimentation) onto the discrete space D_1 (bed number and bed thickness). Under the hypothesis s_2 (constant sedimentation rate) the same set maps onto an alternative discrete space. As turned out during the earlier discussion, there exists a map I which transforms D_2 into D_1, but only if D_2 is constructed from infinitesimally small sampling intervals, i.e. if D_2 contains all bedding planes. Otherwise, if the sampling distance becomes sufficiently large, D_1 cannot be constructed in all details from D_2. On the other hand, from D_1 we can construct all possible outcomes of the model and sampling strategy $s_2(n)$ (n for the number of sampling points). The reason is that P can be reconstructed in all details from D_1 but not from D_2. Thus, diagrammatically

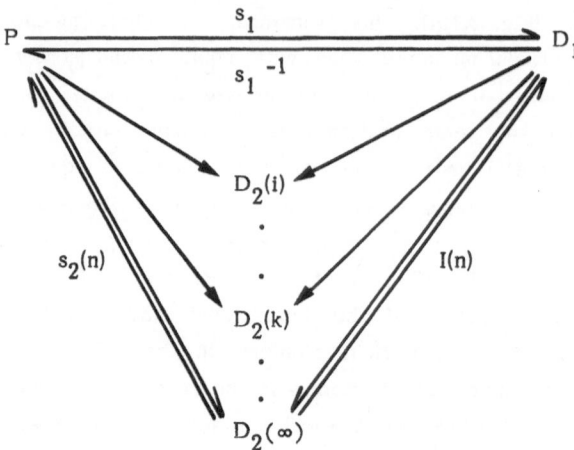

and it turns out that only D_1 is a generic representation of the data, i.e. mathematical analysis justifies the way via geological reasoning. In retrospection, the problems concerning Markov chains are the same -- geologically and mathematically.

3.4   CENTROID CLUSTER STRATEGIES -- CHAOS ON FINITE POINT SETS

A wide field of geological and paleontological research focuses on classification problems. A set of objects -- samples, specimens etc. -- has to be classified in such a way that the elements of a cluster show a maximum of 'similarity' while different clusters have a maximal dissimilarity. As the numbers of objects and variables become large, it is convenient to use computer procedures to solve the cluster pattern recognition problems. The classical approach into this direction became known as 'numerical taxonomy' (SOKAL & SNEATH, 1964). Nowadays, there exists a large number of algorithms which try to solve the pattern recognition problem in a straight forward way by use of various similarity and distance measurements (e.g. HARTIGAN, 1975; STEINHAUSEN & LANGER, 1977; VOGEL, 1975). Clustering strategies produce, in general, no unique solution; an alternative distance measurement or even a different input sequence of the data can change the local and global structure of the clusters (VOGEL, 1975). In addition, our opinion about the image of the statistical universe can fairly diverge from the similarity structure which is generated by a clustering strategy. There are three points to be discussed: the representation of clusters in binary trees, image concepts, and stability problems with cluster strategies. The binary trees imply that they give a classification, but the comparison with classification trees shows that this is not the case. The image concept of most cluster strategies is far from our geometrical intuition.

It will turn out that this discrepancy explains much of the instable behavior -- because the clusters are not geometrical objects, they cannot be used for a true classification, and any additional element destroys the local structure. Besides the instability between clusters, there appears an instability within clusters, which is more striking because it does not depend on some metadefinition of a cluster and leads to true chaotic behavior on finite point sets.

3.4.1 Binary Trees

The cluster structure found by some algorithm is usually represented as a binary tree, which illustrates the similarities or distances between the hierarchies of elements and clusters. The binary trees in cluster analysis are just a picture of one possible similarity structure (VOGEL, 1975); they are not at all a classification by some ordering sequence. Thus, they do not allow to insert an additional object without changing the structure, at least locally, nor do they allow to search an element by some decision rule, as it is possible with binary trees which represent a relation or ordering between data. Binary trees are commonly used as classification structures that allow to search and to insert elements very quickly in sophisticated computer programs (WIRTH, 1975; DENERT & FRANK, 1977), for a mathematical discussion see e.g. SCHREIDER (1975). The difference between binary trees and cluster trees can be illustrated in the following way:

classification trees:

(I)

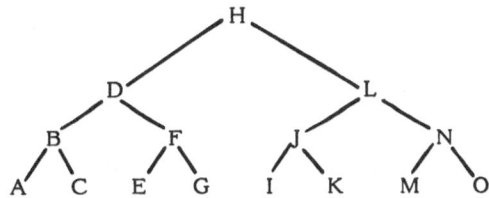

(II)

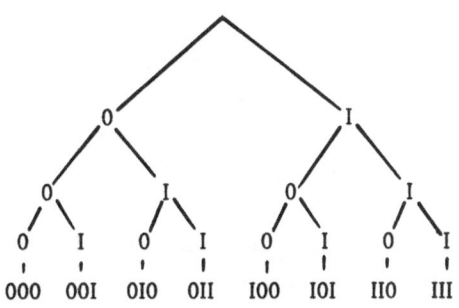

cluster tree:

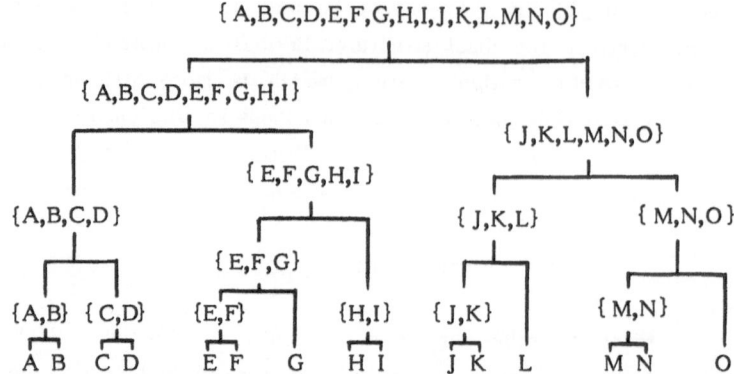

The classification trees represent an ordering relationship like 'before' in tree (I): $A < B < C < ... < O$. Therefore, it provides an optimal strategy to search elements. In the same way, objects can be classified by a sequence of binary decisions 'to have or to have not a certain property'. Tree (II) gives an example for such a classification of the binary numbers. This tree allows to find elements by the decision rules at the nodes, and it allows to insert new elements at its proper position, e.g. it is no problem to insert a number with four digits. Tree (I) provides a minimal classification while tree (II) is an optimal dynamic classification because it allows to insert new elements without any alteration of the present tree structure.

In contrary, the cluster tree does not describe how to arrive at any one of the elements if one starts at the highest hierarchical level. As far as some distance measurement exists between A, B, ..., O, the cluster tree simply states that A is more similar to B than to C. But because the binary tree cannot describe a higher dimensional variable space, it does also not ensure that the distance between A and B may not be identical with the distance between A and C. This property will later turn out to be the main source for instable clustering within clusters. The cluster tree only gives a possible structure of similarities and a possible grouping into subsets -- and not at all a rule how to distinguish the subsets.

3.4.2   Image Concepts

"Cluster analysis is one of the Pattern Recognition techniques and should be appreciated as such" DIDAY & SIMON (1980) write, and

"the general idea of all these methods is to build an identification function (clustering) from two types of information.

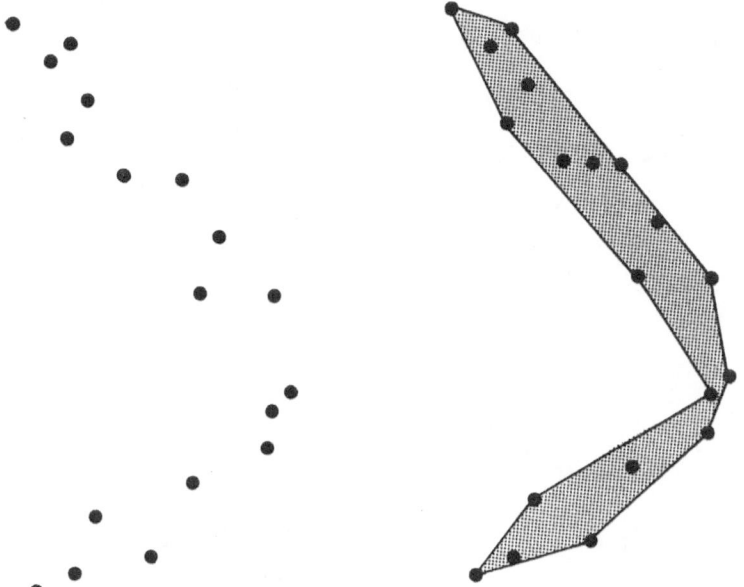

Fig. 3.16: A point pattern in the plane (left) and the intuitive geometrical inter-pretation as a bounded object (right).

> *- on one hand, the experimental result,*
> *- on the other, the a priori representation of a class or*
>   *cluster. "*

Although the concepts of clustering analysis can be well described in a formal mathematical notation (DIDAY & SIMON, 1980; HARTIGAN, 1975), these concepts have some 'weak' points. As Diday & Simon summarize:

> *"In fact these (the cluster) representations rely heavily on the concept of potential or inertial functions. The 'classifier specialist' selects this representation from the knowledge of the properties of his experimental universe."*

Indeed, to work properly with such techniques does not only require to understand the mathematical formulae, but also to understand the qualitative behavior of a cluster strategy and the concept of 'clusters' underlying a certain clustering process.

Given a point set in the plane (Fig. 3.16), our intention will be, in general, that these points are the image of a two-dimensional object with a well defined boundary. In a more general sense, one expects that the points are a sample from a n-dimensional density distribution which is continuously defined in R^n, and, therefore, one suspects

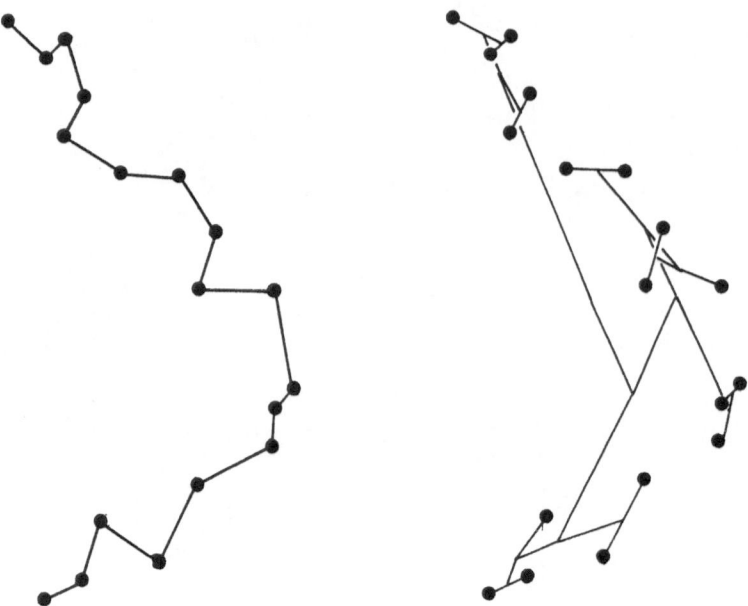

Fig. 3.17: Alternative pattern concepts for the point set of Fig. 3.16 -- a line
(left) and the binary fusion structure of a weighted centroid method (right).

that an increasing sample size gives a statistical universe which can be bounded by
smooth probabilistic surfaces. A consequence is that one identifies automatically 'cluster
recognition' with classification. Indeed, if one has found a cluster and has bounded it
by a closed surface, then any additional data point is accepted as a member of the
point set if it is located inside the bounding hypersurface. A point outside of this 'object'
will be associated with it if its inclusion does not disturb our concept of the image
too much. Thus, the geometrical objects. of our opinion provide classification rules, and
these are possible because the geometrical objects are stable.

Besides this intuitive concept, there are other possibilities to define 'clusters'.
The points of Fig. 3.16 can be connected by a line; the resulting image approximates
a two-dimensional curve (Fig. 3.17). The resulting 'object' can be accepted, but it is
not the kind of things one expects in a two-dimensional space. The binary fusion pattern
of a centroid cluster method (Fig. 3.17) diverges still further from our opinion. That
these pattern concepts appear somewhat strange, may have its reason in the feeling
that they are not stable, and, in fact, they are not stable in a structural sense. Any
additional point alters these patterns, at least locally, in an unpredictable way. Furthermore, these objects generate no classification for additional points, there is no relation
like 'inside' or 'outside', like 'belonging to' etc. Classificatory objects for points in R^n

require that they are bounded by a surface. Therefore, an object in R^3 must be three-- dimensional itself. Points, lines and open surfaces in R^3 are degenerated objects which can exist neither physically nor statistically, they are objects which cannot 'contain' something. The definition of a classificatory object in a discrete n-dimensional variable space requires, therefore, at least n+1 data points or samples. These n+1 points then describe just one object, a triangle in R^2, a tetrahedron in R^3, a hypertetrahedron in R^n. A cluster and classification concept requires that the number of elements to be classified exceeds clearly the number of variables defining the position of the elements in a hyperspace. There should not be more than (n DIV m) objects -- n: number of elements, m: number of variables, DIV: division of integers.

In the applications of cluster analysis it is commonly the case that the number of variables equals, or even exceeds the number of elements or samples. The only structure one can find under such conditions is an ordering by some similarity rule, but one cannot expect that this provides a classification, as it is usually done. In a statistical sense, 'classificatory objects' are closely related to the convex hull of a finite point set (EFRON, 1965). In a topological sense, they are just higher-dimensional analogues of a smooth closed surface, a manifold. Because the point sets are finite, the continuous differentiable structures are replaced by simplices and their unions, polygons and polyhedra.

3.4.3   Stability Problems with Centroid Clustering

Within the various types of cluster strategies, hierarchical methods are the ones most commonly used, and within this subgroup the centroid methods are most popular. The term 'centroid' is here used in a very general sense. It simply expresses that two fused elements are further represented by a single element, the centroid of the original points. It is known from a respectable number of cluster strategies that the constructed distance tree depends, to some extent, on the initial ordering of the input data (VOGEL, 1975). As Vogel found by extensive testing with one special data set, the centroid methods become particularly sensitive to the ordering of the data if an entropy distance measurement is used. Cluster strategies with other distance measurements showed this sensitivity only on the lower hierarchical levels. In palecology one has very nice homogeneous data structures -- the frequencies of species within a sample. In this special case, the usual metric distance measurements make not much sense while the entropy measurement provides a 'natural' distance measurement for these frequency data of species (BAYER, 1982). This distance measurement has, in addition, the property that it approximates the Chi^2-test for homogeneity of the samples (DIDAY & SIMON, 1980); homogeneity of the samples being of special interest in palecological studies. Therefore, a program

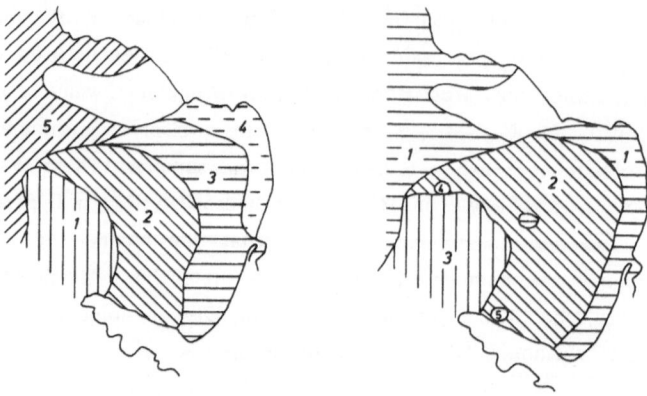

Fig. 3.18: Classification and distribution of the foraminifera fauna of Todos Santos Bay. Left: Classification by hand (WALTON, 1955); right: Ordering by a cluster program (KAESSLER, 1966).

was implemented that allowed to analyze faunal data by the 'entropy method' (BAYER, 1982). The data of a 'hand made' ecological analysis (WALTON, 1955), which had later been reevaluated by cluster analysis (KAESSLER, 1966), were re-reevaluated again to see how the strategy works. Fig. 3.18 gives the spatial faunal distribution pattern of foraminifera, which had been found by the previous workers. Already the first few runs with the 'entropy analysis' showed a strong dependence of the result on the ordering of the input data. The nice statistical properties of the entropy distance measurement, therefore, were useless. But the use of other distance measurements with representation of the clusters by their centroids did not either stabilize the pattern, in contrast to VOGEL's (1975) claim. In fact, dependence from the initial ordering of the data was nearly the same. Fig. 3.19 gives some examples of various tree structures, which result from slightly different initial conditions and parameter settings. The observation that it is not the choice of a special distance measurement, that causes the instabilities, was the starting point for a more detailed analysis of the stability properties of the centroid cluster strategies. This stability analysis will now be outlined.

The widespread use of centroid cluster strategies is, in part, due to their property to produce nice binary hierarchical structures from every data set. The points found to belong to a cluster are replaced by their centroid, which can be either weighted or unweighted. 'Weighted' means that the centroid is computed from all data points within a cluster while 'unweighted' means that a computed centroid is further treated like a data point, or that it has equal weight as a data point. In both cases, the progress of clustering causes a concentration of the original data points into fewer and fewer representatives, the centroids. Thus, every single fusion of two points evacuates locally the point space. As this concentration of the points proceeds in a certain area, it forces

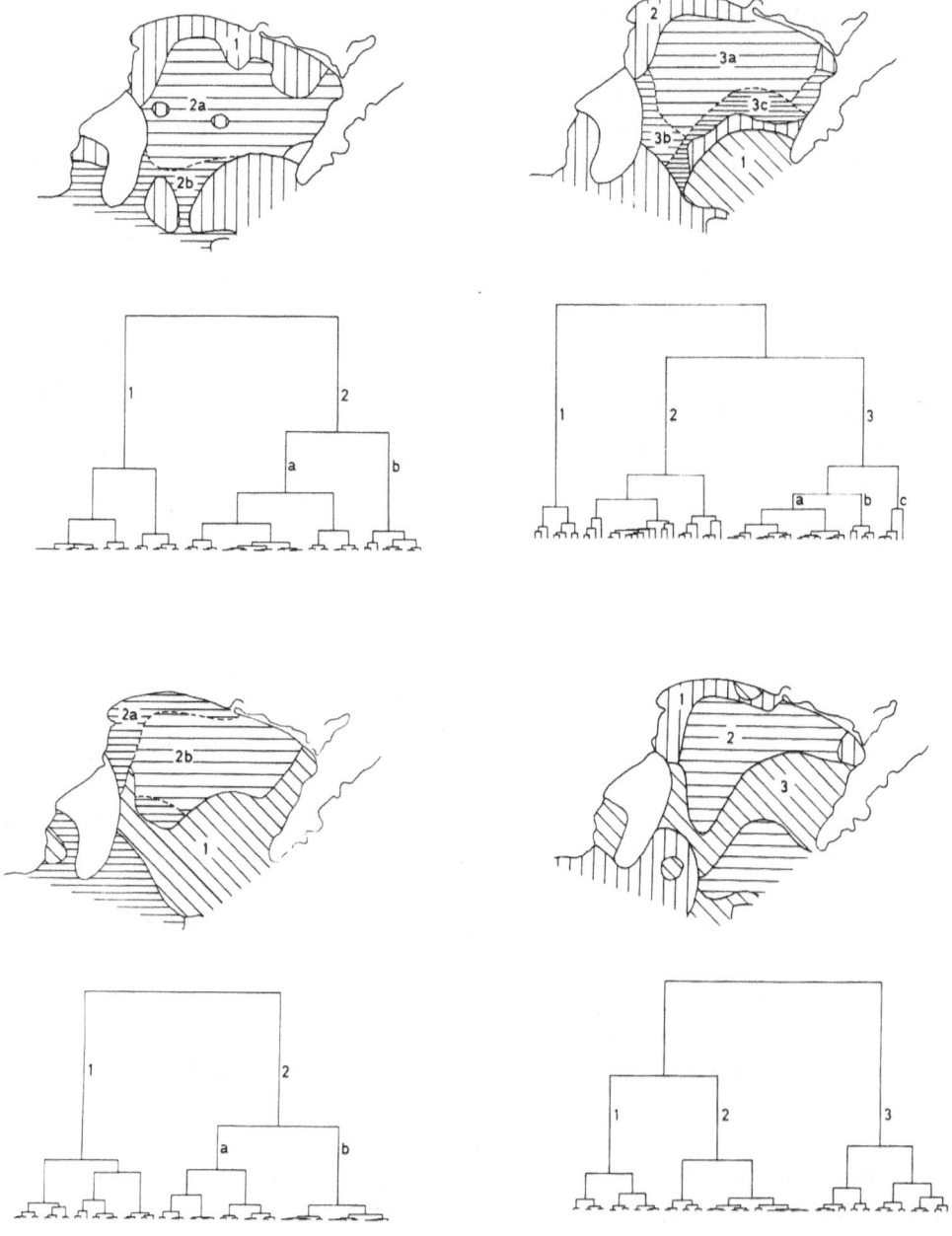

Fig. 3.19: Various classifications with centroid cluster strategies of WALTON's (1955) foraminifera data. Lower left and upper left: Two of a large number of possible cluster trees which result only of an altered input sequence of the data. Upper right: altered metric. Lower right: different clustering algorithm.

the process to branch into other areas which are still more densely covered by data points, and the same process is repeated. This coupled process of local evacuation and subsequent branching into another area of the data space is the reason why every data set is mapped onto a nice binary tree. The tree structure is due to the clustering process, and it is not a property of the data set. The same process causes, in addition, that the distances between the clusters increase monotonically, a celebrated property of these strategies (STEINHAUSER & LANGER, 1977) -- but an alternative viewpoint is that these strategies impose a binary tree pattern onto every data set.

Instabilities between clusters are related to the problem which images or which point configurations can be accepted as a cluster, i.e. one needs a metalanguage definition of a cluster. Although the properties of clusters and partitions, their homogeneity, are usually defined in terms of the metric, the idea of a 'good' cluster is mainly a geometrical one. The question 'what is a good cluster' is commonly reduced to the problem how distant two clusters must be to be separable and whether concave clusters are allowed or not. In clustering analysis the terms convex and concave are replaced by homogeneity and chain-homogeneity (DIDAY & SIMON, 1980). At the moment, we may use the earlier discussed geometrical image concept to define 'good' clusters. Then, concavity means simply that every line connecting two points of a cluster is bound to the interior of an hyperpolyhedron, that forms the boundary of the cluster. Consequently the centroid is not necessarily bound to the interior of a concave cluster (Fig. 3.20). And because every fusion of two points increases locally the distances, the process can reach a state where the image breaks into disconnected pieces, and where these pieces are connected with data points which form separate clusters under the geometrical image concept (Fig. 3.20). The striking point is that in such cases the centroid method does not follow the 'rule of nearest neighborhood'; not the clusters, which appear closest because their boundaries are closest, are fused, but the more distant clusters are connected (Fig. 3.20). This behavior elucidates again that such a pattern representation cannot work as a classification, at least not as a 'natural classification'; in some respects, this pattern concept is even contradictory to our understanding of similarity.

So far, a metadescription of clusters was necessary. One can also simply accept that a cluster formed by a centroid process has nothing in common with our geometrical imagination. If it can be uniquely constructed by an algorithm, it may be a totally abstract structure -- but why should it not be as real as geometry? That clustering does not produce such a well defined abstract pattern, will be shown by a discussion of instabilities within clusters. To avoid any assumption about the structure of clusters, one has to restrict the attention to local structures as they are defined by the clustering process itself. Then one can analyze how a local disturbance propagates into the established hierarchy. Fig. 3.21 gives two cluster patterns over a given point set (compare

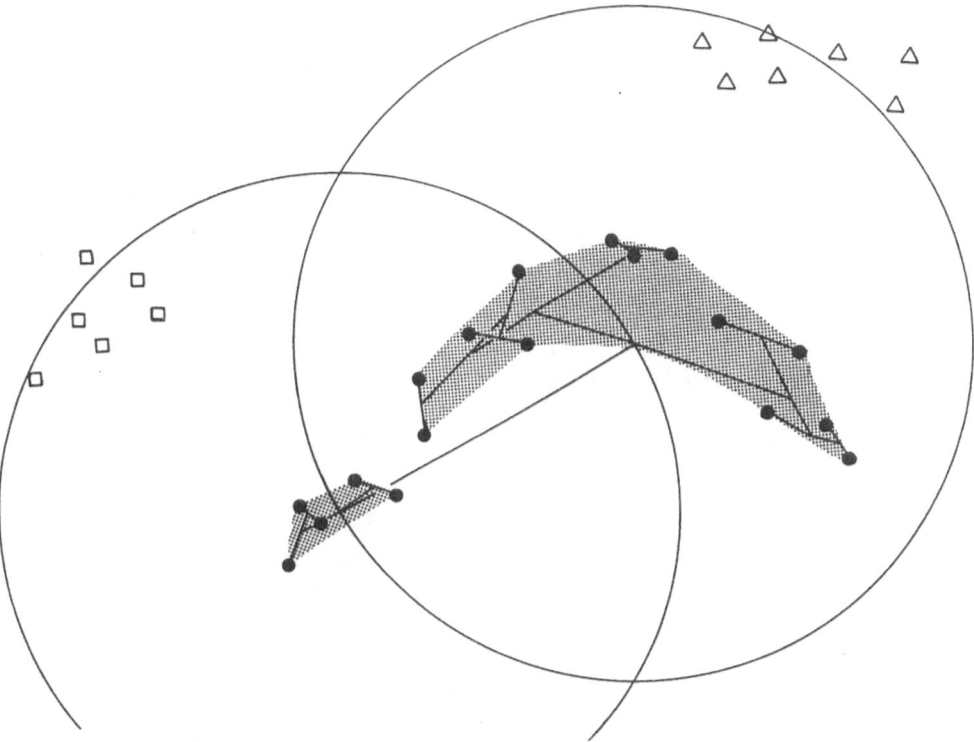

Fig. 3.20: Branching between clusters. The smaller cluster will be fused with the 'rectangular points' because the formation of centroids changed the local topology of the point space. Consequently the larger cluster will be fused with the 'triangular points' in a later step.

Figs. 3.17- 3.18). This time, an unweighted centroid method was used to find the similarity structure. This method causes a local decision problem because there are two choices for the same point to be fused, two neighbors have equal distance. Dependent on this choice, two different local patterns are generated (Fig. 3.21). These local alternatives project onto the global tree structure changing the distances or similarities all the way through the cluster tree (Fig. 3.21). The difference is strong enough to change the overall tree pattern. Thus, one of the two trees appears more compact, and, therefore, it will be more easily accepted as a larger cluster if the trees are embedded in a more complex cluster tree. On the lower cluster level the substructure appears also more compact in the right tree of Fig. 3.21 while in the left one two clusters can still be distinguished on the lower level. The decision, which one of the two clusters will be produced, depends only on the ordering of the data. The computer algorithm has either to take the first pair of data points with smallest distance or the last one for fusion -- a consequence of the binary tree representation. Once made the decision, the local topology of the point pattern has changed, and, therefore, the local decision projects onto all later fusions contacting this area. Thus, a true bifurcation of the solution occurs, and it depends

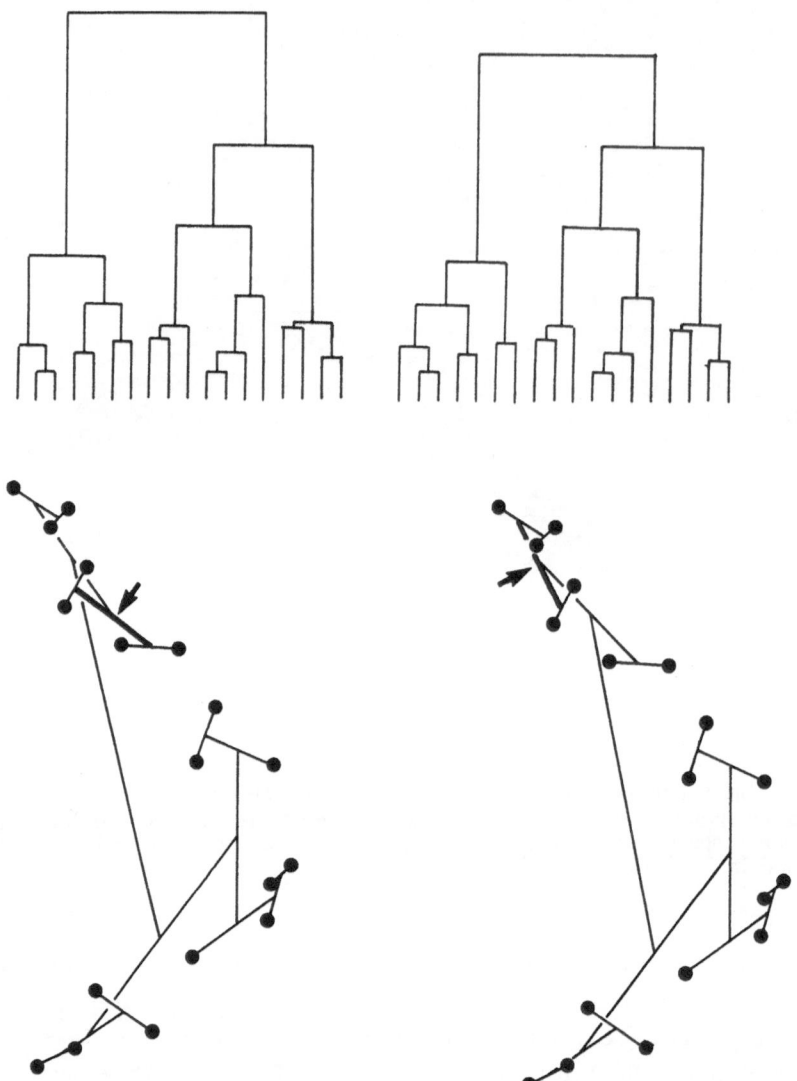

Fig. 3.21: A single local decision problem (arrows) between two equally distant points alters the local tree structure and the distances within the entire hierarchy.

only on the initial conditions (ordering of the data) which branch of the solution will be taken.

The next question is whether there exist situations that cause cascades of bifurcations. In this case, the clustering process would become chaotic, as far as this is possible on a finite point set. Such a sequence of bifurcations can be easily constructed by just

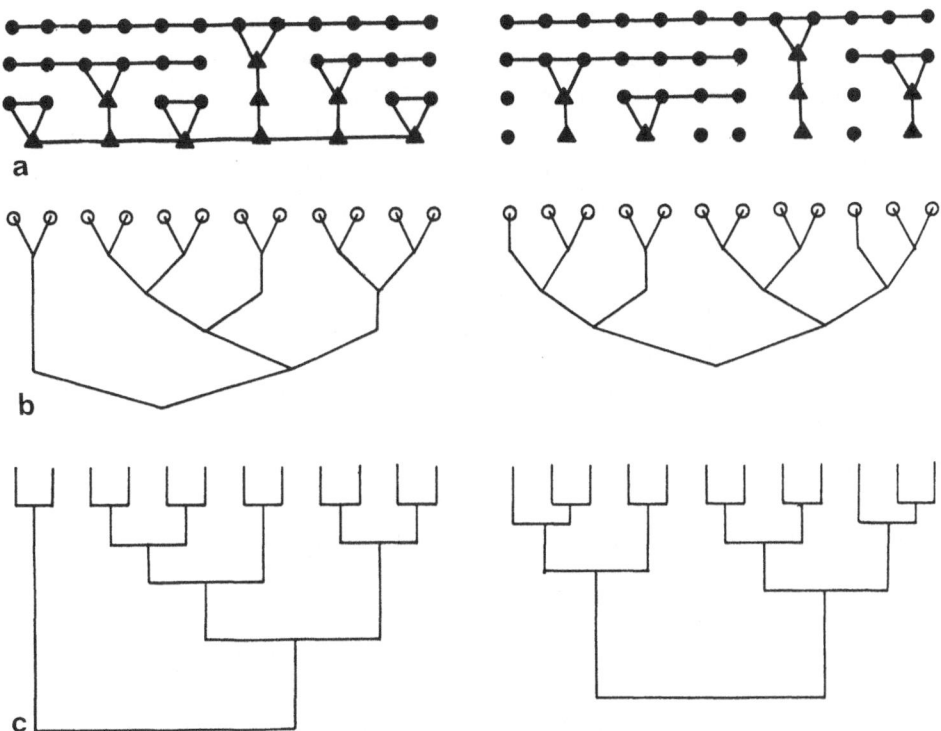

Fig. 3.22: Centroid clustering as a decision game. The figure gives two possible cluster solutions for the identical, equally spaced point set. a) Every fusion causes the bifurcation into two identical subsystems; b) the associated fusion tree; c) the cluster trees.

choosing the most degenerated case of input data -- equally spaced data points. It does not matter whether these points are arranged on a straight line, on a circle or on a regular (hyper-)grid. Equally spaced data points on a straight line provide the most simple case; Fig. 3.22 illustrates what happens in this special case from various viewpoints. First, the clustering process has a dynamical aspect -- at the same time, only two points can be fused, and the fusion of these two points alters the local distance structure between the points. In the case of Fig. 3.22, the initial distance of the equally spaced points, Δx, is altered to $(3\Delta x)/2$ between the centroid and its neighbors. Therefore, the data space is divided into two subsets of points which have still the original structure. But the two substructures are now independent; they can be treated as parallel processes of identical behavior. This bifurcation process proceeds until there are no data points left with the original spacing. Within the limit of a finite point space, any fusion on one of the subsets generates new identical subsets, every bifurcation of the dynamical system generates two dynamical systems of the identical type. In addition, any fusion is a random choice between several possibilities, and the decision, which pair of data points is taken, depends only on the initial data configuration. One has a perfectly chaotic system on a finite point set. Because the point set is finite, the process terminates finally. There are, for every subsystem, two main possible outcomes which depend on some initial choices of the fusion process (Fig. 3.22). One possibility is that during time all points are fused into pairs. Then the resulting centroids are again equally spaced (Fig. 3.22 left), only with altered distances. Thus, the process is cyclic. The alternative is that already the initial fusions determine, to some extent, the further progress because there remain some points of smallest distance on every level which dominate the process (Fig. 3.22 right). The centroid clustering techniques, therefore, provide excellent examples of chaotic behavior on finite point sets. In the case of equally spaced data, every slight change of the input data will cause another cluster tree, i.e. the system is extremely sensitive to the initial conditions. Because the number of data points is finite, the number of possible outcomes is finite as well, it is the number of disjunct permutations of the data. Therefore, the chaotic aspect of the clustering process will increase with the number of (equally spaced) data, i.e. one has the same situation as with the Chi^2-test against a uniform distribution or the trajectories of particles in Galton's machine.

Now, a practitioner could argue that the chaotic behavior of the centroid cluster strategies requires a very special and degenerated data structure. On the other hand, it is well known that most cluster strategies depend to some extent on the input sequence of the data (VOGEL, 1975), and it seems likely that this phenomenon is related to the discussed instability. The remaining question, therefore, is how this instability can arise in a statistical sample. There are two structures which can enforce the chaotic behavior with any kind of data. The first one is that the data are measured with a certain precision. Thus, with increasing number of data one will get an increasing number of equally distant data. Because multiple values at the same point do not change the local topology if they are fused under the centroid condition, the data pattern tends, at least locally, towards equally spaced data. And once more we have the situation that an increase of data does not stabilize the process, as one should expect from statistical reasoning, but that it destabilizes the clustering process. In addition, once more one finds that a uniform or an equally spaced distribution leads to instability. This structure probably is the main reason why the palecological data cause particularly instable clustering. The frequencies are natural numbers which enforce equal spacing locally. The second structure, which can enforce chaotic behavior, is a numerical one. Every computation with a finite number of digits causes numerical rounding or truncation errors. An excellent example due to WIRTH (1972) is the computation of the sum $\sum_{i=1}^{10000} 1/i$, which takes different values if it is computed forward and backward -- the numerical error affects, e.g., 3 of 12 valid digits of a 48-bit machine. The same problem occurs for the computation of distances between data points (and centroids). If the number of variables is high enough, one has to expect that the distance between two points is not symmetric, e.g. in the case of the Euclidian distance one has

111

$$\sum (X_{1i}-X_{2i})^2 \neq \sum (X_{2i}-X_{1i})^2 \; .$$

Therefore, this error will also depend on the input sequence of the data, and this time the error increases with the number of variables and with the complexity of the distance measurement. This can explain why the 'entropy distance measurement' is especially instable. Logarithms are necessary to compute this distance measurement, and this enforces numerical errors.

Thus, the centroid cluster methods turn out to be highly instable due to their pattern recognition concept. They behave in a chaotic manner with respect to slight changes of the initial conditions, namely the ordering of the input data. The chaotic behavior is triggered by the tendency of measured data to form, at least locally, regular spacings, and by the numeric error of the distance measurement, mainly by its asymmetry -- whereby the computation order depends again on the input sequence of the data. In totality, it turned out during the last two chapters that most operations on finite point sets with the computer have to be handled with special care. The local methods, which dominate this field, are usually extremely instable due to small changes of the boundary conditions and of the initial conditions. What we, in general, would need, are rather more rigid topological methods than the highly sophisticated numerical procedures.

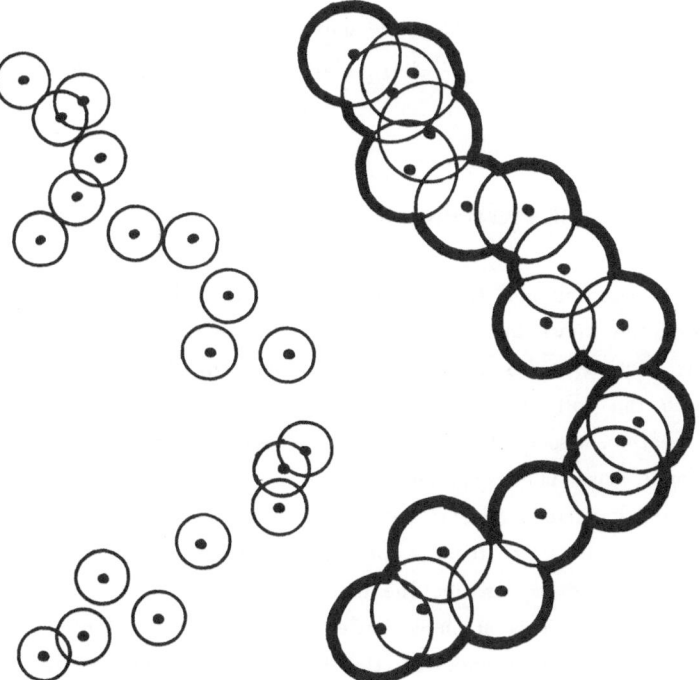

Fig. 3.23: Classification by probabilistic neighborhoods. Points with overlapping neighborhoods belong to a cluster with a well defined boundary (right).

Such a -- probably more -- rigid method was proposed by GRENANDER (1981). The idea is similar to the single linkage methods; however, it has a more geometric background, and this returns to the discussion in section 3.4.2. A probabilistic neighborhood is associated with every point: e.g., circles in the plane, spheres, hyperspheres or alternatively polyhedrons (Fig. 3.23). Every point has a certain probability to be found elsewhere in this neighborhood, and a possible assumption about the probability is 'that the curvature at the boundary is proportional to the density of the probability measure,' GRENANDER (1981), or the probability decreases, as the diameter of the probabilistic neighborhood increases. Two points are grouped in a similarity cluster if their probabilistic neighborhoods overlap. If the radius e of the neighborhood is fixed to a certain value, then a unique partition of the data set arises, and we get partitions of different roughness for different values of e. The resulting structure is not a tree because two or more elements may fuse at once for a certain value of e. However, the resulting 'clusters' have now well defined boundaries as illustrated in Fig. 3.23. The point set of a 'cluster' is bounded by a 'tubular neighborhood', and 'clusters' are different on a certain e-level if their boundaries do not overlap. Thus, given such a classification, we can later add samples and question whether they belong to a 'cluster' for a certain e-value. Clearly, the classification is not stable in the sense that an additional object may cause two or more 'clusters' to fuse without changing the e-level. However, that is a quite more different instability than those discussed above. The classification certainly stabilizes as the number of data points increases and approaches the universe. This approach returns to the examples of the second chapter and is closely related to the continuation problem discussed there.

3.5 TREE PATTERNS BETWEEN CHAOS AND ORDER

The previous presentation focussed at practical problems which arise from unstable algorithms. Here, the viewpoint is much more theoretical, and the questions are mainly conceptual. The section serves as a kind of summary of the previous chapters and connects them with the following one. The morphology of tree-like structures will be the topic of this section, especially the connection between branching pattern and overall shape, which may arise under certain conditions: If one closely studies the crown of a tree, the branching pattern appears rather irregular while an entire view of the crown, from some distance, is characteristic on the species-level (Fig. 3.24). On this basis HONDA (1971) attempted to describe the multifarious form of trees by a few parameters of the branching pattern. This attempt, however, was not new: Already D'Arcy Thompson discussed the cymose inflorescence of the botanists and noticed that these botanical patterns are 'analogous in a curious and instructive way to the equiangular spiral' (D'ARCY THOMPSON, 1952); and, as G.L. Steucek informed me, branching patterns of trees were already studied by Leonardo da Vinci.

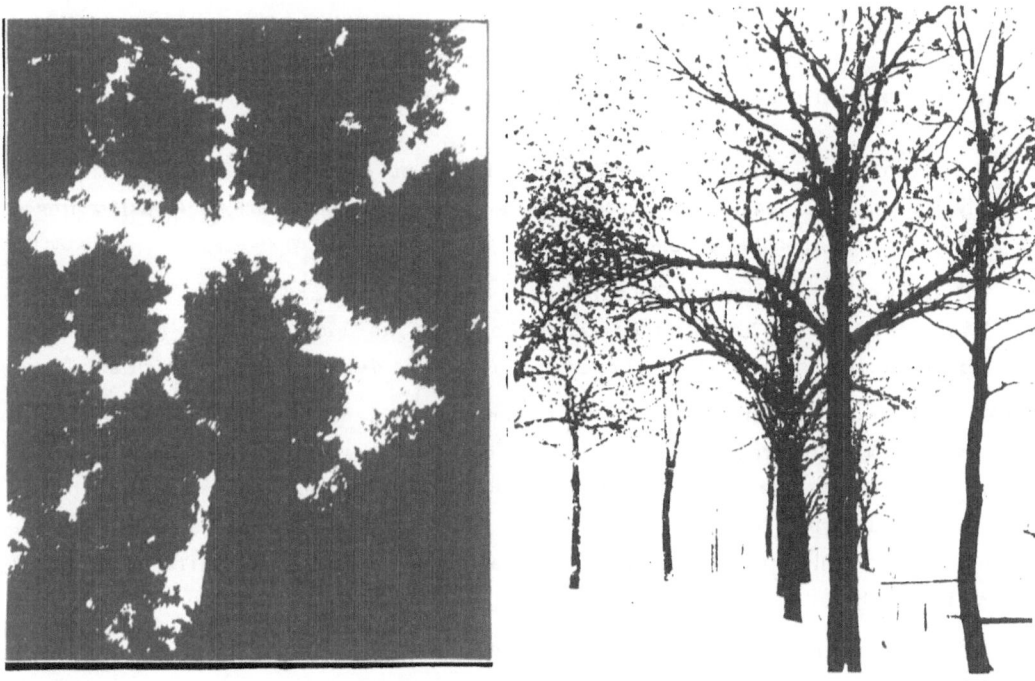

Fig. 3.24: Tree images.

Rather similar branching patterns occur in geomorphology, the network of drainage systems (Fig. 3.25). Trees and networks are equivalent on a certain level, e.g. a tree generates a network if two branches, which come into contact, fuse rather than over-lap -- a situation which necessarily occurs if the branching process is bounded to a

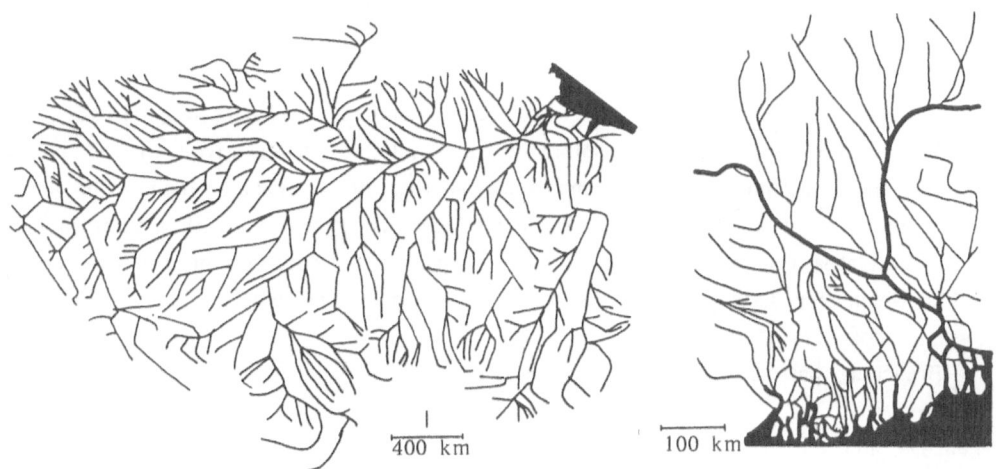

Fig. 3.25: The simplified drainage nets of the Amazon (left) and the Ganges delta (right).

surface or is restricted to a nearly two-dimensional object. The network in leaves provides an example for the latter case. As a concession to the literature, the term 'trees' will be equivocally used as the term 'open networks'.

The botanical attempts toward a description of branching patterns in trees are mainly deterministic and have, therefore, be criticized. Thus NIKLAS (1982) writes in a study on plant branching simulations:

> *"Unless the extent of apparent order that can arise from random processes can be determined, there is no valid basis for asserting that a particular structure implies deterministic causes."*

The meaning of 'random', however, is rather vague. If we attribute a probability to a growing system to branch or to branch not during a growth step, we may alternatively formulate a deterministic system which branches at the beginning of every growth step, but may be disturbed by some external cause which inhibits branching, and such external events may be truly stochastic. This aspect was clearly expressed by OSTER & GUCKEN-HEIMER (1976):

> *"If a series of ... censes are collected ..., and they appear chaotic, exhibiting no perceivable regularities, then we conclude one of three things:*
>
> *(a) the system is truly stochastic--dominated by random influences;*
>
> *(b) experimental error is of such a magnitude that all regularities are obscured;*
>
> *(c) a very simple deterministic mechanism is operating, but is obscured "* *(by the phenomenon described here in section 3.1.1).*

This is a viewpoint which now seems to be common in physics (e.g. HAKEN, 1981; THOMPSON, 1982). Examples of points (b) and (c) have been given in the previous sections, and experimentally it seems more appropriate to exclude points (b) and (c) than to prove point (a); the failure of the Chi^2-test in connection with a uniform distribution illustrates this point: There is still enough to be detected on the deterministic level in which stochastic elements may enter in terms of disturbances.

Branching patterns of tree-like bodies are of some interest because they are not restricted to trees, they occur commonly in biology and paleontology, on the level of organisms and on the level of organs. Besides this, they are even interesting objects in geology, geomorphology, and physics. The discussion of metric trees is, to some extent, based on a study by Bayer & McGhee (An Analytic Approach to Branching Systems,

115

preprint 1984), a study which was initialized by a paper given by G.L. Steucek at Tübingen in 1984.

3.5.1 Topological Properties of Open Network Patterns

In geomorphological studies open networks and tree patterns naturally originate
at the sources of streams, a viewpoint which is opposite to the botanist's one where
the branches usually originate at the trunk. However, in terms of morphometry botanists
take a viewpoint identical to the geomorphological one. The streams or branches are
ordered by the descriptive "stream number", which is due to Strahler (e.g. BARKER
et al., 1973; GRENANDER, 1976):

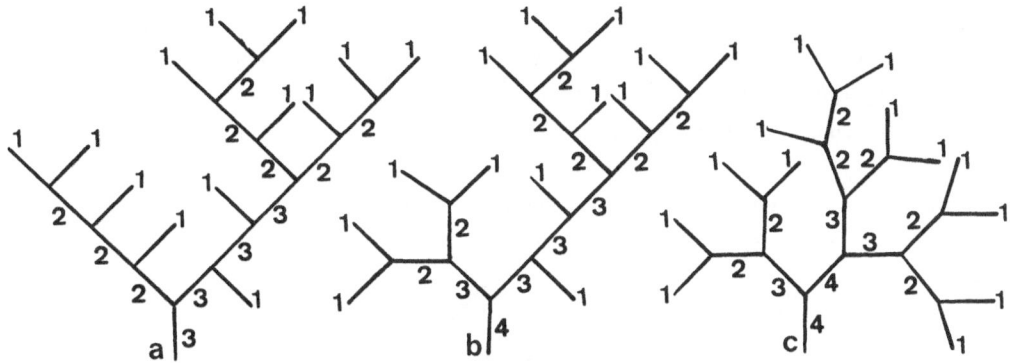

Fig. 3.26: Trees with branches ordered by Strahler numbers. All trees have equal
numbers of end branches.

Definition 1:   The S t r a h l e r   n u m b e r   's',   or the order of a
single branch, is given by the following rules:

(1)   end branches have order s=1
(2)   two branches of order s produce a branch of order s+1
(3)   two branches of different order (s_{max}, s_{min}) produce a branch of order s_{max}.

The highest Strahler number S involved in an open network is called the
o r d e r   o f   t h e   n e t w o r k .

The Strahler number (Fig. 3.26) of branching systems has special properties (e.g. BARKER
et al., 1973; GRENANDER, 1976). In part they are based on empirical observations,
in part they are deducible from the definition. Thus, if the numbers of branches of
each order are counted, then the logarithm of the number of branches in each order
plotted against the Strahler number gives a linear plot: i.e. $\log n_s = a+bs$; where n_s:
the number of branches of order s (BARKER et al., 1973); or, an observation from

geomorphology (HORTON, 1945), the bifurcation ratios are approximately constant. Let again n_s be the number of branches of order s, then the b i f u r c a t i o n r a t i o

$$n_{s-1}/n_s = R_s \qquad (3.34)$$

is approximately constant, i.e the numbers n_{s-1} and n_s are in a geometrical relationship. These 'empirical laws', however, are not unexpected, they result from the definition of the Strahler number. By definition every network of order s must contain at least two branches of order s-1, and the same holds for any substructure in the tree with s > 1. Therefore,

$$n_s - 2n_{s-1} \geq 0 \quad \text{or} \quad n_{s-1}/n_s \geq 1/2. \qquad (3.35)$$

In the case $n_{s-1} = 1/2$, the bifurcation tree is a perfect symmetric tree. The 'empirical laws' are expected whenever the deviation from the binary tree is random, i.e. whenever branches are randomly suppressed and added with a zero-mean.

Another measurement, which is sometimes useful, is the number of end branches of a tree or a branch (GRENANDER, 1976):

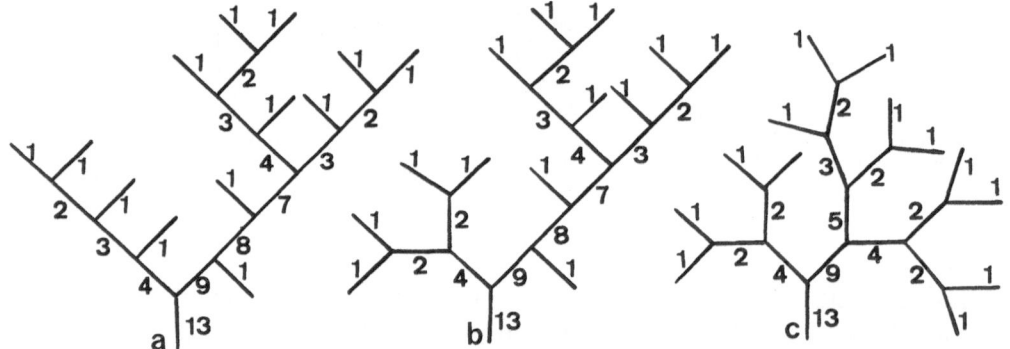

Fig. 3.27: The trees of Fig. 3.26 with branches enumerated by 'numbers of complexity'.

Definition 2: The n u m b e r o f c o m p l e x i t y N_c of a branch of order s is the number of its end branches.

Fig. 3.27 illustrates the characterization of trees by their numbers of complexity. The number of end branches and the order of the network are not independent. The relationship between the number of complexity and the Strahler number is simply

$$N_c \geq 2^{S-1}, \qquad (3.36)$$

and the equality holds only if the tree is a perfect symmetric tree. To prove this relationship it is helpful to introduce

Definition 3: The s k e l e t o n of an open network is its largest symmetric substructure.

The skeleton is constructed by eliminating the branches with Strahler number s_{min} at nodes where $s_{L,i-1} \neq s_{R,i-1}$.

The remaining tree contains only nodes where the Strahler number increases and, therefore, is perfectly symmetric. Along a pathway from an end branch to the trunk the

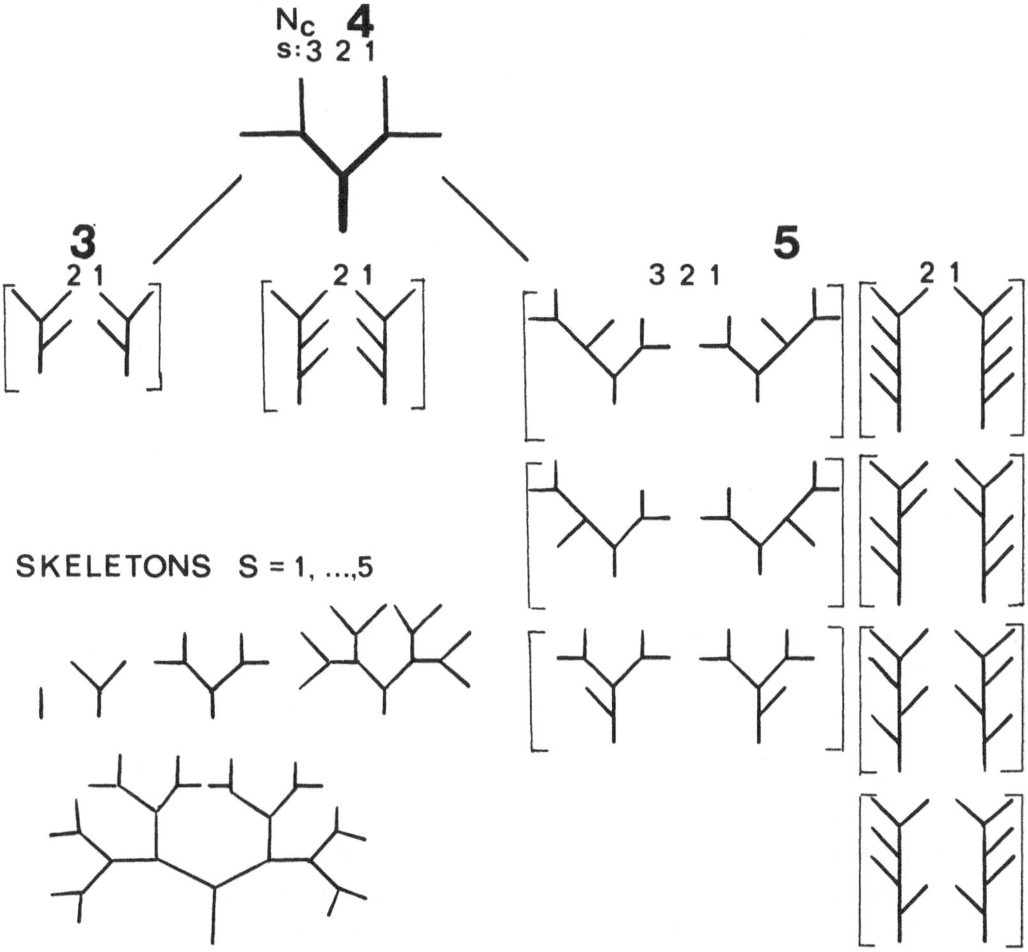

Fig. 3.28: Trees of complexity Nc = 3,4, and 5 (s:Strahler numbers) with distinct permutations of branch patterns. Note that these permutations are not all topologically distinct as indicated by brackets. Lower left: elementary skeletons of Strahler numbers 1 to 5.

Strahler number simply counts the number of bifurcations. If we invert this pathway, the number of branches of equal Strahler number increases monotonously like 2^0, 2^1, 2^2, ... , 2^{S-s}, ... Arriving at an end branch the number of end branches is $N_c = 2^{S-1}$. Relation (3.36) now is easily proved: If the tree is symmetric, then the equality holds as was shown, otherwise branches have been eliminated during the construction of the skeleton, and, therefore, end branches have been eliminated, i.e. the inequality in equation (3.36) holds. This discussion provides us further with a topological interpretation of the Strahler number: It uniquely defines the e l e m e n t a r y   s k e l e t o n   of a tree or its substructures (Fig. 3.28), where 'elementary' means that the length of branches is not considered (empty nodes of the skeleton).

Another question is how many different trees with identical number of end branches do exist, or generally: Given a certain number of end branches, how many distinct networks can be generated? GRENANDER (1976) gives the following solution, which is due to SHREVE (1966):

The number of open networks $N(n)$ for any branching system with n end branchings is

$$N(n) = (2n-2)! \, (n! \, (n-1)!)^{-1}, \tag{3.37}$$

and the number $N(n_1, n_2, ..., n_{S-1}, 1)$ of topological distinct networks of order S with n_1, ... , $n_{S-1}, 1$ branches of order 1 .. S is

$$N(n_1, n_2, \, ... \, n_{S-1}, 1) = \prod_{s=1}^{S-1} 2^{(n_s - 2n_{s+1})} \, ((n_s - 2)! \, ((n_s - 2n_{s+1})! \, (2n_{s+1} - 2)!)^{-1}. \tag{3.38}$$

These two equations provide us with a probability measurement of finding an open network with Strahler numbers $n_1, n_2, \, ... \, , n_{S-1}, 1$; when the number of end branches is given, the discrete probabilities are

$$p(n_1, n_2, ..., n_{S-1}, 1) = N(n_1, n_2, ..., n_{S-1}, 1) \, / \, N(n). \tag{3.39}$$

Fig. 3.28 illustrates the networks which correspond to $N(3)=2$, $N(4)=3$, $N(5)=14$; i.e. the various combinations of branches. However, these trees are not all topologically distinct, some of them are simple symmetric inflections and cannot be distinguished: Let the trees be the two-dimensional projection of a three-dimensional image, then the inflections mean simply that an observer walked around the object by 180° and then classified it differently because the symmetries have changed. The number of topological distinct trees is only

$$N_T = N(n) \text{ div } 2 + N(n) \text{ mod } 2. \tag{3.40}$$

The difference of the two concepts is not trivial, e.g. the probabilities will be different under the two hypotheses: The probabilities for a tree with four end branches (Fig. 3.28) are $p(3,2,1) = N(3,2,1)/N(4) = 1/3$ and $p(2,1) = 2/3$. Based on topological similarity, however, we find the probabilities $p_T(3,2,1) = p_T(2,1) = 1/2$, and the geometrical hypothesis influences the conclusions we may draw from the probabilities.

Finally it seems worthwhile to emphasize that the probabilistic aspect, mentioned briefly, is based on the comparison of an observed pattern with a totally deterministic pattern, the symmetric tree -- a possible question would be: How strong do we have to disturb a binary tree to arrive at the observed tree pattern. The concept of skeletons can be used to emphasize this point. If the symmetric trees without repetitions of empty nodes -- the elementary skeletons -- are defined as primitives of our tree structures, then any tree can be defined as 'product' of elementary skeletons: A node is the product (Si,Sj), where the first term means left, the second right branching (for symmetry reasons this notation is arbitrary) and Si defines an elementary skeleton of order (Strahler number) s=i. The trees of Fig. 3.26 can be described as lists (Fig. 3.29)

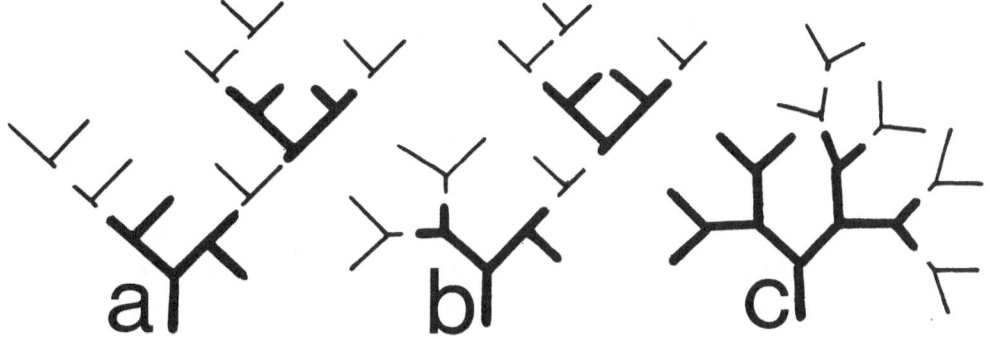

Fig. 3.29: The trees of Fig. 3.26 decomposed into elementary skeletons.

(a) S3: ((S2: (S2,O), O, (S2: (O, S3: ((S1,S2), O, O, S2), O)
    or   ((S2,O), O , (O, (S1,S2), 2*O, S2), O)

(b) S3: ((S2, S2, S2: (O, S3: ((S1,S2), O, O, S2)), O)
    or 2*S2, (O, (S1,S2), 2*O, S2), O)

(c) S4: (O, O, O, O, S2: (O,S2), S2, S2, S2)
    or (4*O, (O,S2), 3*S2)

where the brackets contain descriptions of the end branches clockwise; the term

'O' indicates termination of the branch, and '*' means repetitions. The description is recursive and, of course, could be reduced to binary decisions or s-expressions which form the basic structure of the programming language LISP. Elementary LISP and the lambda-calculus provide another conceptual framework to handle such structures (e.g. DENERT & FRANK, 1977).

3.5.2 Pattern Generators for Open Networks

The theme of the previous section was the description of tree patterns. A common problem, however, is the simulation of tree patterns and the question how many different patterns can be generated from some simple rules. Essentially, there are two approaches: transformations of qualitative properties and metric models. Two examples will be discussed here. In both cases, the process envolves some primitive elements and deterministic production rules. However, one can adapt probabilities to the bifurcation events and then simulate randomly disturbed patterns; such systems were e.g. used by Raup to analyze whether phylogenetic trees are random structures (RAUP et al., 1973).

A) Algebraic Models -- Prototypes of Branching Patterns

One class of models, which generates regular structures inclusively tree patterns, is based on the transformation of 'alphabets': A string of symbols like (a,b,c,...) characterizes qualitative properties of an object, and a set of transformation rules like (a → ad, b → c, ...) defines the evolution of these properties, for instance how they transform during time. Such systems have been extensively applied to developmental systems by LINDENMAYER (1975). The theory of such transformation groups is much older and closely related to the theory of permutations; for a detailed discussion see ASHBY (1956, 1974). A typical system, which produces tree patterns, is (LINDEN-MAYER, 1975):

> the alphabet or set of primitives: (a,b,c,d,e)
> the production rules: (a → db, b → c, c → d, d → e, e → a),
> starting with the element 'a' generates the tree

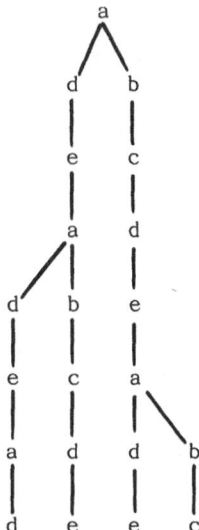

Inspecting the tree one finds that it consists of the regular repetitions of strings (d,e,a) and (b,c). The redefinition A = (d,e,a), B = (b,c) and the modified production rules (A → AB, B → A) simplify the tree pattern

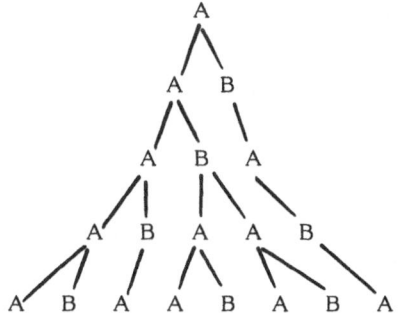

which, however, contains the same information as the original tree because we can restore it by the inverse substitution. The two trees are equivalent, and we can produce more equivalent trees by substitutions of the type $A=(a_1, a_2, \ldots, a_n)$, $B=(b_1, b_2, \ldots, b_n)$, where the sets of primitives in A and B are disjunct, i.e. do not contain identical elements. Two trees can now be defined as equivalent if it is possible to reduce them to the same primitive form; the reductions, which are allowed, are those which do not alter the singular pattern of the production rules, i.e. which preserve

termination   points like A → A,

bifurcation points A → BC,

and loops A → B → A.

The problems, which arise if other reductions are used, were briefly mentioned during the discussion of Markov chains in sedimentology.

The reduction process may be illustrated by a somewhat more complicated system:

the alphabet: (a,b,c,d,e,f,g,h,i,k)

the production rules: (a → bc, b → kd, c → lk, d → gb, e → cf, f → ih,
g → hi, h→ de, i → k, k → k),

which are rather complicated and generate an evenly complicated pattern

a

bc

kdek

kgbcfk

khikdekihk

kdek kgbcfk kdek

kgbefk khikdekihk kgbcfk

khikdekihk kdek kgbcfk kdek khikdekihk

The system was used by LINDENMAYER (1975) to describe the structure of compound leaves. The strings of symbols were interpreted as cells along the margin of the leaf which posses certain morphological properties. We try to reduce the complicated production rules and observe first that the element 'a' occurs only once; indeed, it serves only as an entry point. Regular repetitions of substrings in the tree occur from the third production downward, these are A=(kdek), B=(kgbcfk), C=(khikdekihk), and the production rules can be simplified (A → B, B → C, C → ABA)

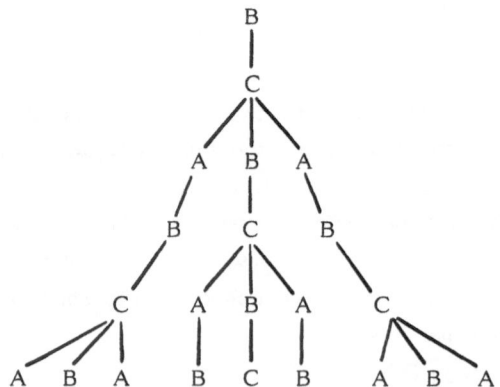

The reduced system produces a nice tree with triple points at C, a structure which was not obvious from the original production rules, but, clearly, the reduced system is equivalent to the original one -- if we substitute the original values for A,B, and C, we have the original productions.

Inspecting the elements A,B,C it turns out that they all are of the form (k ... k), i.e. the 'k'-elements act as brackets of the original strings. Only in the element C, k's occur within the string. If we want that this fine structure is not lost in the reduced version, we can introduce the radicals $H=(k,h,i)$ and $H^T=(i,h,k)$ replacing the element C, which now reads $C=HAH^T$, and the productions are

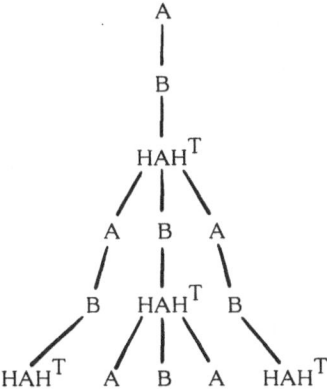

which again comprise the structure of the original rules.

This section attempts to illustrate that the reduction of complicated systems to equivalent simple ones is an essential step in pattern recognition problems. However, such reductions can only involve those transformations and redefinitions which do not affect the singular structure of the original system. Using the concept of equivalence and proper reduction processes may also save much investigation time because a rather complicated system may turn into a well known simple one, and the classification of such structures reduces to a (not necessarily finite) catalogue of prototypes.

B) A Metric Model -- the Honda Tree

HONDA's (1971) approach was to formulate a pattern-generator which allows to study the form of trees by means of the computer, based on a minimal set of parameters (Fig. 3.30). Here only his 'geometrical assumptions' are of importance, they are repeated in a slightly modified form:

I   The branches are always straight, and their girth is not considered;

II   the mother branch produces two doughter branches with each branching event;

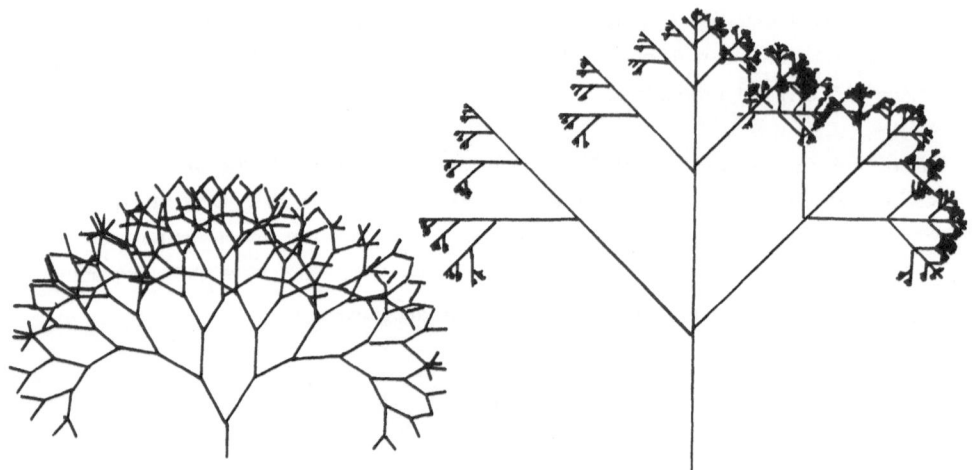

Fig. 3.30: Two-dimensional images of Honda trees in the (x,y)-plane. The right figure has in part been completed to approximate tripling at branching points. Still more realistic bifurcation patterns for plants can be constructed if the branches fork off under a 'divergence angle' which follows a Fibonacci series (cf. HONDA, 1971).

III   the length of each successive doughter branch is shortened in a certain ratio with respect to the mother branch;

IV   the doughter branches form constant angles with their mother branch throughout the tree; however, left and right branching angles may be different;

V   the mother branch forks off into doughter branches in the plane which contains the mother branch and whose steepest gradient line coincides with the direction of the mother branch.

From these assumptions Honda derived a formula which allows to compute the successive coordinates of branching points and, thus, can be programmed for graphical computer output. Point (V), which is not a separate point on Honda's list, is responsible for a rather complicated structure of Honda's formula, which has to be applied separately for left and right branches (different parameters R and θ):

$$x = x_B + R(u \cos\theta - (Lv \sin\theta)/(u^2+v^2)^{1/2} \qquad (3.41)$$

$$y = y_B + R(v \cos\theta + (Lu \sin\theta)/(u^2+v^2)^{1/2}$$

$$z = z_B + R w \cos\theta$$

where R: ratio between subsequent branches, θ: branching angle and $u = x_B - x_A$, $v = y_B - y_A$, $w = z_B - z_A$, $L = (u^2+v^2+w^2)^{1/2}$.

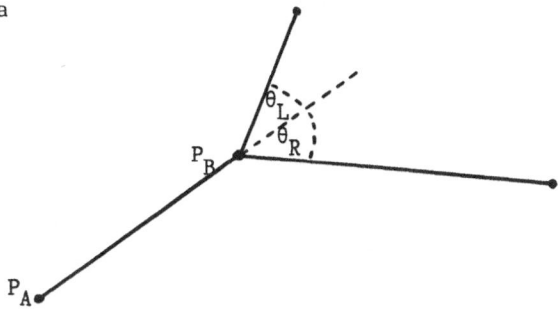

Fig. 3.31: Honda's notation of a bifurcation.

x_A, x_B etc. are coordinates of the branching points (P) as indicated in Fig. 3.31. Honda's formula is easily programmed and evaluated with a computer; however, equation (3.41) is not of the proper form for an analytic analysis; it takes a much more convenient form if one does not work in global coordinates but considers the dislocations of branching points which are

$$\begin{bmatrix} x-x_B \\ y-y_b \\ z-z_b \end{bmatrix} = R \begin{bmatrix} \cos\theta & -b\sin\theta & 0 \\ b\sin\theta & \cos\theta & 0 \\ 0 & 0 & \cos\theta \end{bmatrix} \begin{bmatrix} x_B-x_A \\ y_B-y_A \\ z_B-z_A \end{bmatrix} \qquad (3.42)$$

where $\quad b = L(u^2+v^2)^{1/2} = (1 + w^2/(u^2+v^2)^{1/2}.$

The matrix, which appears on the right side, describes a rotation and a simple compression (or dilatation). We can separate the two components by dividing all matrix elements by a factor

$$R_1 = (\cos^2\theta + b^2\sin^2\theta)^{1/2} = (1 + w^2(u^2+v^2)^{1/2}\sin^2\theta)^{1/2}),$$

and the dislocations can be written

$$\begin{bmatrix} u_{i+1} \\ v_{i+1} \\ w_{it1} \end{bmatrix} = R_1 R \begin{bmatrix} \cos\theta & -\sin\theta & 0 \\ \sin\theta & \cos\theta & 0 \\ 0 & 0 & c \end{bmatrix} \begin{bmatrix} u_i \\ v_i \\ w_i \end{bmatrix} \qquad (3.43)$$

where $\quad \tan = b(\sin\theta)/(\cos\theta)$

and $\quad c = (\cos\theta)/(\cos^2\theta + b^2\sin^2\theta)^{1/2}.$

The factor R_1 defines an additional shortening ratio which depends like the matrix element 'c' on the rotation angle and the length of the mother branch. However, only the length of the mother branch is variable (cf. assumption IV). The relevant coefficient,

Fig. 3.32: Vector notation of a bifurcation.

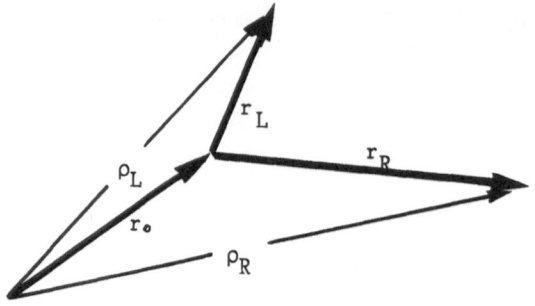

which depends on this parameter, is $w^2/(u^2+v^2)$, cf. equations (3.42) and (3.43). In case this quotient is constant, equation (3.42) is a simple linear system. We shall not discuss Honda's original model in more detail; however, condition (II) can only be satisfied if $w^2/(u^2+v^2)$=constant. Only in this case, the number R defines the shortening ratio of branches, and Honda's model can be replaced by equation (3.43).

Honda's model provides a framework which captures essential aspects of branching patterns in tree-like bodies although already Honda needed some additional assumption for three-dimensional trees. However, it can be simplified leading to a description by two sets of vectors: r, the dislocations of branching points, and ρ, the radius vector of the branching points in global coordiantes (Fig. 3.32):

The h o d o g r a p h of a H o n d a t r e e is defined by the pair of iterative equations

$$r_{R,i+1} = a_R A_R r_i \qquad (3.44)$$
$$r_{L,i+1} = a_L A_L r_i \; ,$$

and the space coordinates of the Honda tree are given by

$$\rho_{R,i+1} = \rho_i + r_{R,i+1} \qquad (3.45)$$
$$\rho_{R,i+1} = \rho_i + r_{L,i+1} ;$$

r, ρ will denote vectors throughout this section; R,L are the indices for left and right branching; 'A' is an orthogonal matrix (cf. equation (3.43)); 'a' is a scalar and denotes the ratio between subsequent branches.

While in a Honda tree s. str. $0 < a_R, a_L < 1$, we call a tree with arbitrary values of a_R, a_L a g e n e r a l i z e d H o n d a t r e e, and, where not necessary, the attribute 'generalized' is dropped.

A Honda tree, thus, is a binary tree, and if we superimpose some probability that branches may be lost, it generates an open network, as discussed in the last section, with all its properties. The Honda tree, developed for botanical models, possesses a property which is a classical observation in geomorphology: the constant ratio of branches. The stream length satisfies the approximate relation (GRENANDER, 1976)

$$l_{s-1}/l_s = R_L, \qquad s = 2,3,...,S \qquad (3.46)$$

l: length of branches, s: Strahler number, R_L:length ratio,

an empirical observation, which coincides with Honda's assumption (II). We shall derive some properties of Honda trees, which are only based on this assumption, and, therefore, can be applied to stream patterns.

3.5.3 Morphology of Branches in Honda Trees
"D'Arcy Thompson's Problem"

Honda trees, as defined in the previous section, possess special properties: constant branching ratios and branching angles. These properties allow to compute the length and morphology of certain branches -- under special circumstances even for infinite numbers of branching events. Furthermore, the discrete model will turn out to be isomorphic with a differentiable system.

A) Length of Branches

By definition the branches of a generalized Honda tree are in a geometrical relationship, i.e. the successive length of branches is given by the iterated map

$$| r_i |= a \; |r_{i-1} | \qquad (3.47)$$

where 'a' may be either the branching ratio for left or for right turns, or better a_{max} or a_{min}, a notation used now because left and right are of little meaning (cf. section 3.5.1). If one considers a continuous sequence of turns in one direction, the length of the branch is given by

$$|r_i |= | r_0| \; \Sigma \; a^i, \qquad (3.48)$$

a geometrical series with the well known sum

$$L =| r_0 | \; (1-a^{n+1})(1-a)^{-1}. \qquad (3.49)$$

In case of a convergent Honda tree, the branching ratios have to satisfy $0 < a_{min} < a_{max} <1$, and we can compute the length of such a branch even for $n \rightarrow \infty$:

$$\lim_{n \to \infty} |r_0| (1-a^{n+1})(1-a)^{-1} = |r_0| (1-a)^{-1}.$$

(3.50)

From this observations we establish the

Proposition: In a Honda tree the sums

$$L_{max} = |r_0| (1-a_{max}^{n+1})(1-a_{max})^{-1} \quad \text{and}$$

$$L_{min} = |r_0|(1-a_{min}^{n+1})(1-a_{min})^{-1}$$

provide upper and lower bounds for the length of any pathway through the tree with fixed n (n = 1,2, ... ,k) and initial length r_0 . In case $a_{max} = a_{min}$, all possible pathways are of equal length. In case $0 < a_{min} < a_{max} < 1$, this holds even for $n \to \infty$.

Consider any other pathway, it necessarily includes at least one turn opposite to the continuous path, i.e. an element $a_{max}^{i} a_{min}^{j}$. However, for any combination of i and j we have

$$a_{max}^{i+j} \leq a_{max}^{i} a_{min}^{j} \leq a_{min}^{i+j}.$$

(3.51)

The sum over the length of elements along an arbitrary pathway with initial length r_0 contains only products of this type, and the length is bounded for fixed n because every element is bounded. The equality in the equations holds if $a_{max} = a_{min}$, and independent of the pathway, its length is constant for fixed n.

These results can immediately be applied. Consider a distributive system which has to be optimized in the way that the lengths of pathways with common origin are equal. This optimization problem is solved by any tree with $a_{max} = a_{min}$, independent of its morphology, e.g. branching angles.

If one considers generalized Honda trees, one cannot expect that the length of branches converges for $n \to \infty$, however, there may be special pathways with finite length. To get an idea what branches converge we consider pathways consisting of regular repetitions of 'm' left and 'n' right turns, for which we write

$$(m*R, n*L), (m*R, n*L), \ldots \quad ; m,n \in (0,1,2,\ldots,).$$

The sum of such a series is

$$(L-|r_0|) = \quad a + a^2 + \ldots + a^m \qquad \text{; the first sequence of right turns}$$

$$+ a^m(b + b^2 + \ldots + b^n) \qquad \text{; the first sequence of left turns}$$

$$+ \ldots +$$

$$+ a^{km}b^{(k-1)n}(b + b^2 + \ldots + b^n) \quad \text{; the } k^{th} \text{ and final repetition.}$$

This sum can be rewritten as

$$L - |r_0| = (\sum a^i + a^m \sum b^i)(1 + a^m b^n + (a^m b^n)^2 + \dots + (a^m b^n)^k) \tag{3.52}$$

or

$$L - |r_0| = (\sum a^i + a^n \sum b^i)(a^m b^n)^k / (1 - a^m b^n),$$

and branch length converges if

$$0 < a^m b^n < 1. \tag{3.53}$$

Given a certain a_{max} and a_{min} we have to find numbers m,n so that equation (3.53) holds; if they exist, the branch is of finite length. In some sense, these special pathways provide us again with upper limits: If (m*R, n*L) converges and $a_R > a_L$, then all pathways (m*R, (n+i)*L) converge even if 'i' is some random variable but satisfies i ≤ 0.

B) Branching Angles -- Similarity and Self-Similarity

The second variable in Honda's model is the branching angle. However, this variable is not a proper measurement because the tree pattern depends strongly on the branch ratios. If one wants to compare two trees, a measurement is necessary which takes the branching angles and the branch ratios into consideration. This leads to the following

Definition: The s i m i l a r i t y   i n d i c e s of a Honda tree are the two numbers $S_{max|min}$ = (log $a_{max|min}$)/θ ; θ in radians. Two trees are absolutely similar if both indices are identical, they are partially similar if they agree in one similarity index.

To motivate this definition we consider the hodograph of the Honda tree, i.e. the vectors $r_i = aAr_{i-1}$, and we consider again a continuous pathway which consists entirely of left or right turns. Then the vectors r_i evolve like r_0, aAr_0, $(aA)^2 r_0$, ... $(aA)^n r_0$. Because A is an orthogonal matrix (cf. equation 3.43), the powers $r_n = a^n A^n$ can be written as

$$r_n = a^n \begin{bmatrix} \cos n\,\theta & -\sin n\,\theta & 0 \\ \sin n\,\theta & \cos n\,\theta & 0 \\ 0 & 0 & c^n \end{bmatrix} r_0. \tag{3.54}$$

The simple redefinition ψ = n θ transforms this equation into

$$r_n = e^{c\psi} \begin{bmatrix} \cos\psi & -\sin\psi & 0 \\ \sin\psi & \cos\psi & 0 \\ 0 & 0 & c^n \end{bmatrix} r_0 \qquad (3.55)$$

$$c = (\log a)/\theta$$

that is, the vectors r_n are located on a logarithmic spiral in the (x,y)-plane or on a trochospiral in three dimensions, whereby the term 'spiral' includes circles and straight lines. To see this more clearly consider $r_0 = (x_0 y_0 z_0)$ in its most general form, and equation (3.55) can be rewritten as

$$r = \begin{bmatrix} x_0 & -y_0 & 0 \\ y_0 & x_0 & 0 \\ 0 & 0 & z_0 \end{bmatrix} e^{c\psi} \begin{bmatrix} \cos\psi \\ \sin\psi \\ c^n \end{bmatrix} . \qquad (3.56)$$

The term in brackets is clearly a logarithmic spiral if we allow ψ to vary continuously. Equations (3.54-56) provide a continuous approximation for the location of dislocation vectors in a monotonous sequence of left or right turns, and the similarity indices are based on the similarity of these spirals taking into account that a change of the magnitude of the branching ratio can be compensated by the branching angle so that the vectors are still located on the identical spiral. Another aspect is that a Honda tree is everywhere selfsimilar in terms of its hodograph (Fig. 3.33). Equation (3.56) shows that we can build the hodograph entirely of sequences of continuous left and right turns, i.e. of spirals in the (x,y)-plane. Along such spirals the branching points are defined by $\psi = \theta n$, and they give rise to a spiral in opposite direction. In case the Honda tree is absolutely convergent, the two initial spirals bound the hodograph, i.e. all other dislocation vectors are located inside these leading spirals. The Honda model thus describes an ideal selfsimilar system with possible infinite repetitions.

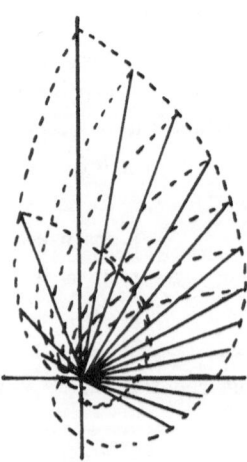

Fig. 3.33: The hodograph of a Honda tree and its continuous approximation.

C) Branches and Bifurcations -- a Quasi-Continuous Approximation

So far, we gathered information about the branching pattern without regard to the form of branches; however, it will turn out that most work was already done. We have seen that the hodograph of a Honda tree consists of similar spirals which originate at a 'leading' spiral, and in the case of a convergent Honda tree they all approach the same coiling center:

> Definition: A l e a d i n g   b r a n c h   is a continuous sequence of either left or right turns, and it is the continuous approximation of a leading branch. The term   's p i r a l'   includes trochospirals and circles and straight lines.

The branches and the bifurcation pattern are found if it is possible to 'integrate' the hodograph, i.e. one has to sum the dislocation vectors along every leading spiral:

$$\rho_i = (\sum_{i=0}^{K} a^i A^i) r_0, \tag{3.57}$$

that is again a geometric series, but this time it involves a matrix. However, if one uses the inverse of a matrix -- denoted by M^{-1} -- this series can be summed like an ordinary power series (e.g. ZURMÜHL, 1964), and the sum takes the form

$$\rho_i = (I - aA)^{-1}(I + a^{n+1}A^{n+1}) r_0 \tag{3.58}$$
$$\text{where I: the identity matrix,}$$

or in a more extensive form

$$\rho_i = (I-aA)^{-1} r_0 - (I-aA)^{-1}(a^{n+1}A^{n+1}) r_0.$$

This equation describes the succession of bifurcation points along our leading spiral if we use 'n' as variable. It turned already out that a term $(aA)^n r_0$ describes a sequence of vectors (cf. equation 3.54) with the vertices located on a spiral, and returning to these arguments again it becomes clear that the variable term in equation (3.58) describes again the spiral of the hodograph. The term $(I-aA)^{-1}$ is constant and can be decomposed into a rotation and an elongation, as illustrated in section 2.5.2 B. The leading spirals of the hodograph, therefore, map onto similar spirals, the branches. The representation of a branch by a sequence of spirals is illustrated in Fig. 3.34. The form of branches, or more precisely, the location of bifurcation points, thus is well defined in a Honda tree and restricted to a single family of functions. The spirals are the same as in the hodograph up to similarity transformations.

The bifurcation pattern thus can be approximated by spirals which originate at

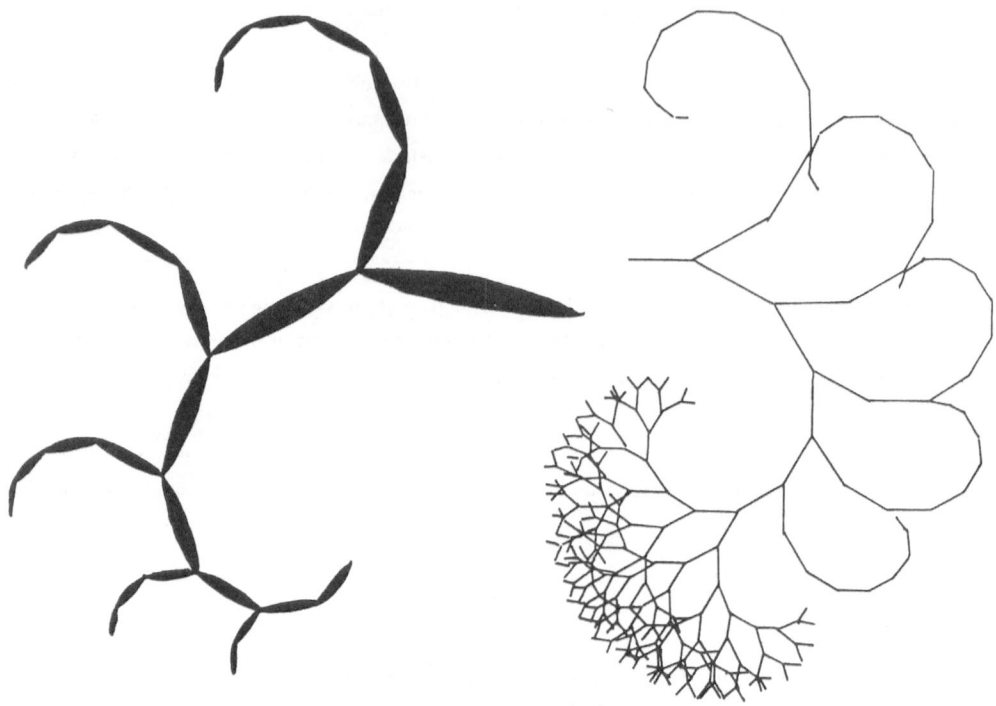

Fig. 3.34: A Honda tree as a sequence of spirals.

a leading spiral and give rise to further spirals and so on. The Honda tree provides a model for

"D'Arcy Thompson's" theorem: A branch system with continuous branching angles and geometric relationships between mother and doughter segments of branches is isomorphic with a system of leading spirals: The first generation of branches is attached to the leading spiral in regular, usually geometrically increasing or decreasing distances measured by arc length and give rise to a second generation, and so on.

This pattern has been illustrated above, here we give some more properties which will be needed later. Thereby we restrict ourself to the two-dimensional image of Honda trees in the (x,y)-plane (cf. equation 3.44):

Propositions concerning the form of branches in Honda trees:

(1) There are at most two different spirals in a bifurcation tree of Honda type. All others are similar to these two spirals. Similar spirals have identical coiling

directions.

(2) All branches of a certain generation are directed to the same side of the leading spiral.

(3) The length of a spiral arc between two branching points is proportional to the length of the discrete branch segment between these points.

(4) The lengths of any two spirals with equal numbers of branching points are in an allometric relationship. If the two spirals belong to the same family of leading spirals, their lengths are simply proportional.

Proposition (1) follows from the fact that a Honda tree contains only bifurcations with constant branching angles and branch ratios. The two spirals are defined by equations (3.54-56). Proposition (2) follows from the definition of a leading spiral which is a continuous pathway of only left or right turns, and from the constancy of branching angles. To prove proposition (3) we choose the coiling center of a leading spiral as origin of the coordinate system and assume the bifurcation points are at regular angular distances Φ=constant. The succession of these points is given by ρ_i=exp(-iϕ). The length of a spiral segment is

$$L_s = c^{-1}(c^2+1)^{1/2}(|\rho_i| - |\rho_{i-1}|) \qquad (3.59)$$
$$= c^{-1}(c^2+1)^{1/2}(1-e^c)e^{-ci}.$$

The length of the discrete branch segment connecting the same bifurcation points is

$$L_b = |\rho_i - \rho_{i-1}| = r_i \qquad (3.60)$$
$$= (e^{-2ci}+e^{-2c(i-1)}-2e^{2ci+c}((cos\phi i)(cos\phi(i-1)) + (sin\phi i)(sin\phi(i-1))$$
$$= (1+e^{2c}-2e^c cos\phi)e^{-ci}.$$

The only variable is 'i', all other terms are constant. Therefore, we can express the length of discrete branch segments in terms of the length of the spiral arcs

$$L_b = a L_s. \qquad (3.61)$$

The length of a branch for a given 'i' is the sum over subsequent segments. The constant factors, however, need not to be summed, the sum involves only terms exp(-ic) which are identical in both expressions. Equation (3.61), therefore, holds for any fixed number of branch segments 'i'. In both equations (3.59+60) the ratio between subsequent segments is $L_{b,i+1}/L_{b,i} = L_{s,i+1}/L_{s,i} = exp(-c)$, that is the geometrical relationship in D'Arcy Thompsons theorem. To prove proposition (4) we start with the allometric relationship between two logarithmic spirals:

$$\rho_1 = a_1 e^{c_1\phi} ; \qquad \rho_2 = a_2 e^{c_2\phi},$$

c_2 can be expressed in terms of c_1 by an equation $c_2=bc_1$ or $\rho_2 = a_2 exp(bc_1)$, and, therefore,

$$\rho_2 = (a_2\rho_1/a_1)^b, \qquad (3.62)$$

i.e. the radii are in an allometric relationship. The length of a logarithmic spiral is proportional to the length of its radii (equation 3.59), and the allometric relationship holds also for the length of spiral arcs for any fixed angular interval iϕ. If the two spirals belong to the same family, the coefficient b=1.

As long as we consider branches of finite length, the results hold for any values of branching ratios. There are some special cases, e.g. if one branching ratio equals

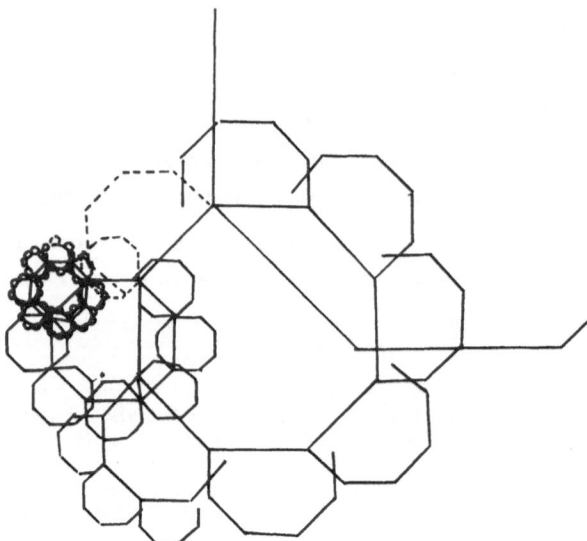

Fig. 3.35: The plane image of a Honda tree if $a_{max}=1$, $0 < a_{min} < 1$ are regular polyhedrons.

one. In this case, the images of branches in the (x,y)-plane are circles as illustrated in Fig. 3.35 for $a_{max}=1$, $a_{min} <1$. If such a special parameter setting is chosen, the net pattern depends strongly on the bifurcation angles, e.g. if $a_{min}=0$, then the system resembles the 'elastic collision in a circle' as discussed in section 3.3.2 (Fig. 3.14). If both branching ratios equal one and the branching angles are $2\pi/3, \pi/2, ...$, the Honda tree degenerates into regular triangular, rectangular and hexagonal grids which cover the entire (x,y)-plane. Usually, however, we will find sequences of logarithmic spirals which, in the case of convergent Honda trees, are bounded in length. This causes the chaotic appearance in the two (and three-) dimensional images of the trees, the end branches cluster and overlap prohibiting the recognition of regular structures.

3.5.4 Evolution of Shape

'Shape' or 'outline' means that an object is bounded by some kind of surface. Simple bifurcation diagrams, e.g. the skeletons S1-S4 in Fig. 3.28, do not possess a shape in this sense. However, as the number of branches increases, an outline evolves, at least subjectively. From previous work we know that a convergent Honda tree needs to have a maximum outline or limiting shape for a sufficiently high number of bifurcations. This limiting shape exists and cannot exceed a certain boundary because all branches are of finite length. On the other hand, the number of end branches increases with the number of bifurcation events like $n=2^s$ (s=Strahler number because the tree is symmetric) and clearly goes to infinity if the number of bifurcation events is not bounded. BARKER et al. (1973) estimated the mean number of buds arising from the highest order branch in a birch tree to be >7000.

As the number of end branches increases, we expect increasing density of end points and a trend towards a continuous outline. Fig. 3.36 elucidates this point where only the end points of the trees of Fig. 3.30 are marked. Clearly, the end points cluster in certain areas and give the illusion of continuous lines which together give the

Fig. 3.36: Distribution of branch endings in the trees of Fig. 3.30.

illusion of a continuous outline. One question to be discussed is whether these points may define a continuous curve or even fill some space, as the number of bifurcation points goes to infinity. The other question is if there exists some defined envelope or hull of a convergent Honda tree which can be considered as its limiting shape.

A) Trees, Peano and Jordan Curves

To illustrate the concept of shape in more detail we consider a quite different system, continuous curves which are nowhere differentiable. A special example, which is useful in this context, is the Koch curve (cf. MANGOLDT-KNOPP, 1968). The closed version of a Koch curve can be constructed in the following way (Fig. 3.37): A regular triangle (equal angles) is inflected and superposed on itself so that the sides are divided into three segments of equal length, the resulting pattern is a regular star. The corners of this star are again regular triangles, similar to the original ones, and with each

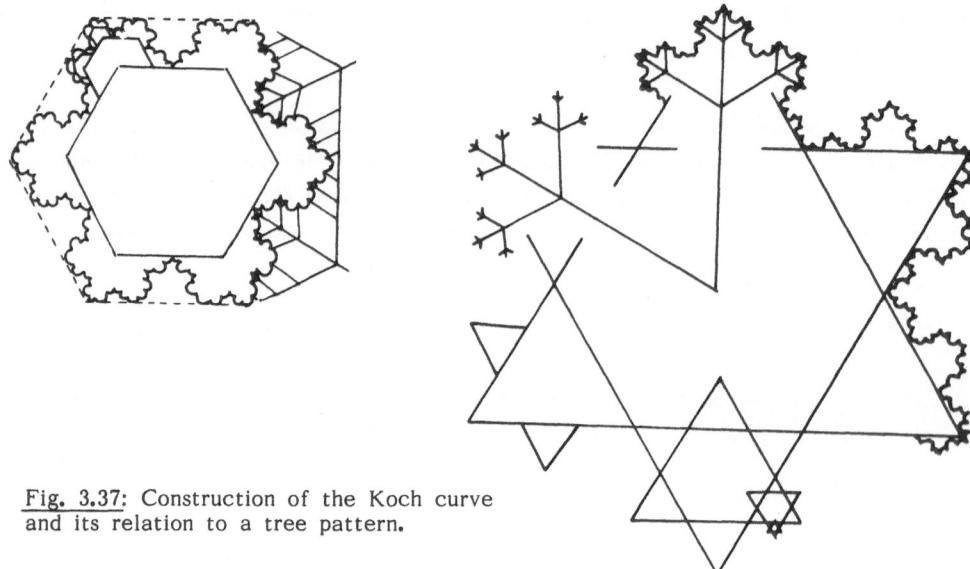

Fig. 3.37: Construction of the Koch curve and its relation to a tree pattern.

of these triangles the process is repeated. The Koch curve is the outer boundary of the resulting pattern as illustrated in Fig. 3.37. If the process is repeated infinitely, the triangles shrink to points and give rise to a continuous, non-differentiable curve. If we now connect the centers of successive triangles by straight lines, these form a tree pattern with a triple point at each branching event. The branching angles and the shortening ratios are constant, i.e. if we delete one of the branch directions, the resulting binary tree is a Honda tree.

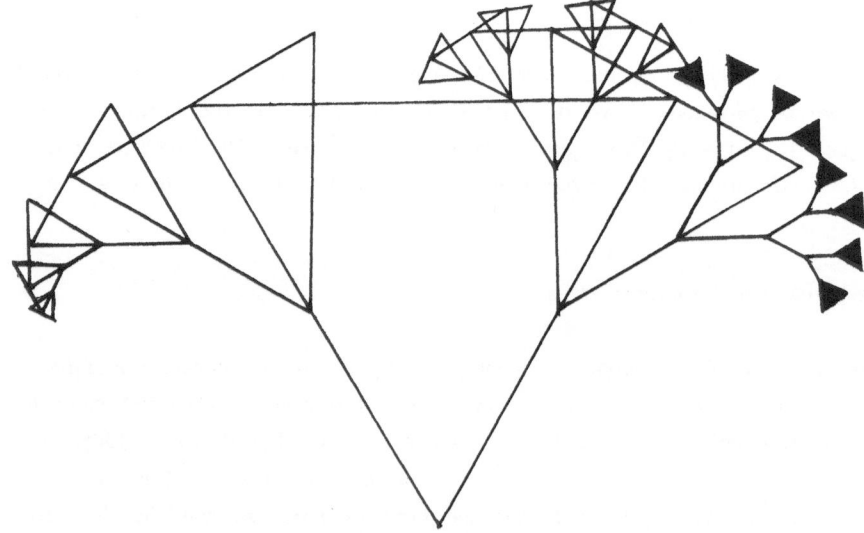

Fig. 3.38: Construction of a binary tree as sequence of similar triangles.

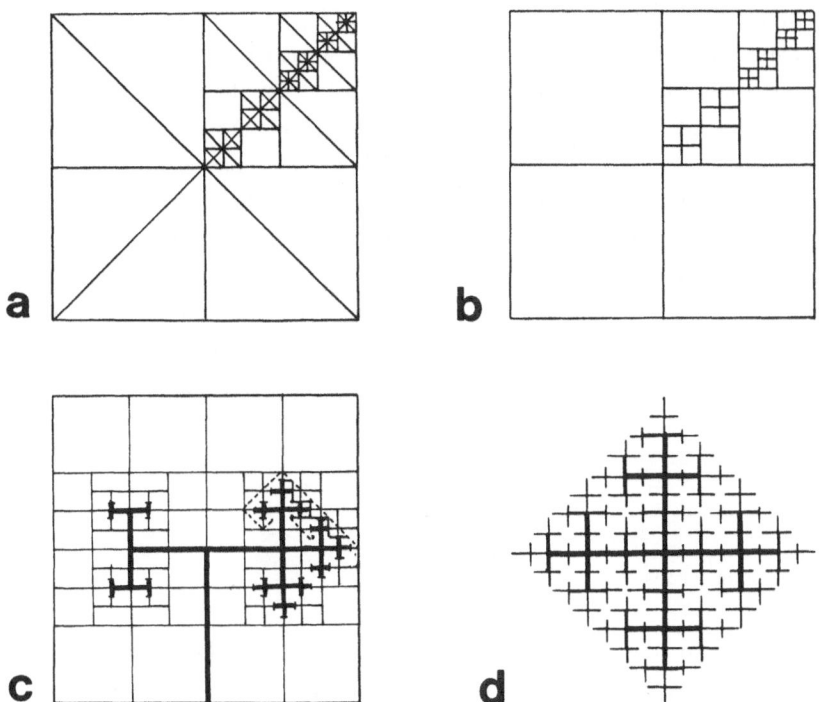

Fig. 3.39: Triangulation of a rectangle by iterative averaging (a; cf. Fig. 2.46) reconsidered as a space filling Peano curve (b). Connection of subsequent centroids generates tree patterns. c: bifurcation and tripling with 'limit shape'; d: space filling tree pattern (see text).

The center of every triangle along the boundary is the termination point of a end branch. As the iteration process goes to infinity, the triangle shrinks to its centroid which, by definition, is the termination point of an end branch, and we expect a continuous curve as limit shape for the tree. In detail, this assumption is not quite correct. If we only consider branches as indicated in Fig. 3.37, then there are no branches terminating in the concave intervals of the Koch curve while the original construction produces triangles in these areas. In the tree pattern an empty interval remains between terminal points, and new ones are generated with every iteration. The points belonging to the tree are only a subset of the Koch curve. If the points, however, are replaced by objects of finite area, the boundary becomes again densely covered. Numerically this has been studied by HONDA & FISHER (1978, 1979) in terms of the most equitable distribution of leaf clusters. Fig. 3.38 shows a similar construction of a binary bifurcation pattern with terminal triangular 'leaves' which tend toward a continuous outline, but again there are areas which are not covered, and in addition the leaves overlap.

The relationship to the discussion of cluster trees is obvious; however, let us return for a moment to a problem of the first chapter, the reconstruction of a continuous surface by iterative averaging (section 2.4.5). If we draw the successive nets in either way illustrated in Fig. 3.39, we get a nested sequence of squares with their centroids as unique limit point. The sequence of centroids can be parametricized (e.g. MANGOLD-KNOPP, 1968; GUGGENHEIMER, 1977), and, as the process goes to infinity, we have a space filling curve or a Peano curve. Again we can connect a sequence of centroids in the way indicated in Fig. 3.39 generating another special tree pattern with its termination points a subset of the Peano curve. As the process goes to infinity, the termination points of a binary tree cluster; however, they are always well separated. The tree becomes more dense if we allow triple points (Fig. 3.39), and finally we can turn the branching process into a space filling process by the condition that branches of any generation sprout into areas of least density (with maximal distance from neighboring branches).

B) The Outline of Honda Trees

The brief excursus to such strange structures as Jordan and Peano curves provides us with two contrasting aspects: It encourages the previous view that there are tree structures which generate an outline, and it discourages the attempt to search a description of this outline by inspecting the distribution of branch terminations. A conclusion could be that we now need statistical methods to study the distribution of branch terminations more deeply; the situation is quite similar to the problem to find a closed boundary for a finite point set as discussed in terms of cluster strategies. However, in the special case of Honda trees we have continuous differentiable functions which approximate the branches, and, what we ar looking for, is another continuous approximation of the most extended outline. Therefore, we replace the discrete model by a continuous approximation hoping to replace the multivarious outputs of a computer program by a theorem about the shapes, which possibly can be constructed from the model under consideration, similar to the model's original attempt to describe the essential structures of complex natural patterns by a few parameters.

Let us consider a leading branch and replace it by its spiral approximation. The spirals of the second generation are then most economically described in local coordinates of the leading spiral, i.e. in terms of its local moving frame or Frenet trihedron. If we consider straight branches of second order, we can e.g. express them in terms of the tangent vector of the leading spiral and rotate this vector into the proper position. The local description of the discrete system is analogous to equation (3.57), only the vector r_0 is replaced by the rotated tangent vector bBt:

$$\rho_i = (\sum_{i=0}^{j} a^i A^i)(bBt). \tag{3.63}$$

If we move this spiral of second order along the leading spiral, the branching angle is everywhere preserved. Consider a convergent Honda tree with infinite number of branching points, then by proposition (4) of the previous section the length of the second order spiral is always in an allometric relationship with the length of the leading spiral, and the same holds for the radii of the two spirals. We take the coiling center of the leading spiral as the origin of the global coordinate system, and the global coordinates of a point fixed on the second generation branching pattern takes the form:

$$\rho_B = \rho_s + a \, |\rho_s|^b \, (bBt) \tag{3.64}$$

where the index s denotes the leading spiral.

Before discussing this equation in more detail, we observe that b=1 if we restrict ourself to leading spirals of the same system because these are all similar. Furthermore, if we consider any fixed point on a spiral of higher generation, this point is described by a vector which originates at the leading spiral. This holds for any point, even for the most distant one. Although it is cumbersome to evaluate the most distant point, it will turn out that we can give a qualitative answer about the outline.

Remaining in a system of leading spirals equation (3.64) is simplified and can be written in a more extensive form as

$$\rho_B = e^{-c\phi} \left[\begin{bmatrix} \cos\phi \\ \sin\phi \end{bmatrix} + \sqrt{\frac{b}{(1+c)^2}} \begin{bmatrix} -c & -1 \\ 1 & -c \end{bmatrix} \right] \begin{bmatrix} \cos\phi \\ \sin\phi \end{bmatrix}$$

or $\tag{3.65}$

$$\rho_B = \left\{ I + bB \begin{bmatrix} -\cos\alpha & -\sin\alpha \\ \sin\alpha & -\cos\alpha \end{bmatrix} \right\} e^{-c} \begin{bmatrix} \cos\phi \\ \sin\phi \end{bmatrix},$$

and once more it turns out that this equation describes the original spiral which is rotated around its coiling center. Every family of leading spirals, therefore, is bounded by a similar spiral -- because a tree consists of two systems of leading spirals, the shape results from superposition of the two systems. On the other hand, within every system of leading spirals there exists an infinity of identical subsystems which differ only in size, and we can consider the maximal outline as result of the superposition of smaller subsystems as illustrated in Fig. 3.40. Concerning the density of termination points we note that the distance of branching points decreases regularly along the leading spiral, the density of branching points is proportional to the curvature of the leading

Fig. 3.40: A Honda tree is self-similar on all levels. Superposition of images on a certain level generates the identical but enlarged image.

spiral, and the same holds for any secondary leading spiral, etc.. Thus, the density increases towards the coiling centers, and the density of coiling centers is proportional to the curvature of the leading spiral. Density distributions of this type were already encountered with the 'cluster strategies'.

Finally we consider branches of finite length. The propositions of the previous section still hold, the only difference is that the leading branches do not coil to infinity. Equation 3.64, therefore, takes the form

$$\rho_B = \rho_s + a(|\rho_s| - |\rho_0|)^b (bBt),$$ (3.66)

and the branches terminate for $|\rho_s| = |\rho_0|$. Far from this singular point and for b=1 the shape is quite close to the spiral of the leading branch. The system, therefore, is really self-similar on each level.

C) Chance and Determinism

The previous discussions showed that the Honda model generates ideal self-similar and deterministic structures: The morphology of branches is (infinitely) repeated and impresses its pattern onto the shape of the entire structure, and the shape is again repeated on the level of every generation of branches differing only in size. With these properties it is unlikely that the Honda model reflects reality -- it is a strongly idealized model, and in a later study HONDA & FISHER (1979) introduced additional sources of variation to approximate real trees more closely. However, self-similarity and fractal systems are current fields of interest (MANDELBROT, 1977), and the Honda tree provides a simple linear model although the basic model is rather unrealistic. However, there is some possibility that models "nearby" approximate reality, i.e. that disturbed Honda trees provide better approximations. There are several ways to disturb the original model, which involve stochastic and deterministic aspects.

A simple approach is 'chance', i.e. some external and perhaps internal noise is added which alters branching ratios, branching angles, and may even inhibit branching. However, 'chance' involves also events as 'to evolve at a certain place', to 'evolve within a certain time interval', or to 'evolve under certain and for the individual structure not predictable environmental conditions', even if they are not random but exhibit only a complex spatio-temporal distribution. The random aspect "if there would not have been that particular thunderstorm at that particular time, then those branches would not have been broken off and those buds would not have been inhibited" is only a very special viewpoint.

Other aspects are external and internal constraints, the system has to fulfill certain boundary conditions and cannot grow freely. A possible result could be that under certain constraints only one or few of the topologically distinct trees discussed in section 3.5.1 are stable. Or, consider the drainage systems of Fig. 3.25; although they are irregular, they are not random; the geology determines much of the drainage system: So the influence of the Andes obviously determines stream directions at the western margin of the Amazon drainage system, and sea-level fluctuations through glaciation periods determined much of the drainage pattern of the Ganges delta. If we consider such systems as random or partially formed by chance, the reason is simply that we cannot reconstruct the historical evolution of boundary conditions to a sufficient precision.

A third group of factors, which may be of interest, are internal constraints, or internal regulation systems. The Honda model is a special case of the family of linear maps

$$x_{i+1} = a_1 x_i + b_1 y_i \qquad y_{i+1} = a_2 x_i + b_2 y_i \qquad (3.67)$$

which is equivalent to the discretization of a pair of coupled linear differential equations

$$dx/dt = a_{11}x + a_{22}y; \qquad dy/dt = a_{21}x + a_{22}y \qquad (3.68)$$

which generate spiral patterns in the phase space for parameter setting equivalent to Honda's model (cf. section 2.2.1). Another possible extension is that the local evolution depends not only on the mother branch (a "Markovian" situation) but on other branches nearby, e.g. that the inhibition of a branch influences growth in some neighborhood. If the resulting regulation is linear, then the map (3.67) comes close to a transport equation as discussed in section 3.1.2. Any local disturbance then would alter the system in some neighborhood. Adding diffusion terms to equation (3.68) and assuming that the coefficients a_{ij} may depend on (x,y) or $|(x,y)|$, leads to the $(\lambda\omega)$-models which play some role in spiral chemical waves (e.g. DUFFY et al., 1980; VIDAL & PACAULT, 1982). Still more complicated systems arise if non-linear feedback mechanisms or autocatalytic systems are introduced (cf. MEINHARD, 1984; NICOLIS & PRIGOGINE, 1977). Such a system is e.g. the Hénon map (cf. THOMPSON, 1982)

$$x_{i+1} = y_i + 1 - ax_i^2, \qquad y_{i+1} = bx_i \qquad (3.69)$$

which exhibits a strange attractor and chaotic behavior. THOMPSON (1982) concluded from the random response of such a deterministic model:

> *"Strange attractors may thus have a profound effect on our modelling of seemingly random behaviour, since it is now seen that a stochastic modelling may no longer be essential in all cases. For simple deterministic mechanical systems, they mean that computer results of their non-linear dynamics must be inspected with care (as must any result of a conventional Krylov--Bogoliubov averaging technique), since one feature of a strange attractor is that a sudden leap in response may occur after a long period of apparent quiescence."*

Another point is also elucidated by this section: Sometimes it is worthwhile to study computer models analytically to detect their internal structure. We usually think the other way, we formulate a problem analytically and then use the computer to solve it. Computer modelling and simulation thus became decoupled from classical analytic mathematics to some extent, providing us usually with an enormous or infinite number of possible solutions. In this respect, nature is rather closely approximated, the data are not summarized, but we can establish an infinite catalogue of forms without the disadvantage to go to the field.

143

A major point throughout this section was self-similarity which leads to the aspect that it is sometimes worthwhile to decompose complex structures into smaller units and to extract the 'primitives' which allow to describe and to analyze the complex global pattern. Fig. 3.41 illustrates this in the case of ammonite sutures which are close to tree patterns, and which are self-similar to some extent. By turning a global problem into a local one the analysis of a structure commonly is simplified -- however, the inverse procedure is not always possible as was elucidated by various examples.

The analytic approach, however, usually condenses the possible patterns and allows to discover unexpected structures which may evolve under certain conditions. Thus, the previous discussion of shape was incomplete, as new patterns may arise if the second order branches are directed to the concave side of the leading spiral. Straight branches e.g. then overlap and are bounded by their evolute, and this pattern does not require the convergence of the Honda tree. Such patterns -- evolutes and caustics -- will be the topic of the next chapter.

Fig. 3.41: Ammonite suture lines are comparable to tree patterns. Commonly they are also self-similar (including reflections) on various levels. The enlarged 'primitives' can repeatedly be found within the complex suture lines.

4. STRUCTURAL STABLE PATTERNS AND ELEMENTARY CATASTROPHES

In the preceding discussion it became clear that most of the observed instabilities are due to the fact that the pattern recognition process lacks an inherent stability property. As LU (1976) states:

"In any branch of science, it is always a challenge to try to classify the objects under study. Unfortunately, it is often extremely difficult to carry out this classification. It becomes much easier if one tries to classify only the stable objects. ... stable objects have boundaries where discontinuities appear. We all know that mathematics used in almost all sciences so far is based on the differential calculus, which presupposes continuity. There is a great demand, therefore, for a mathematical theory to explain and predict (if possible) the occurrence of discontinuous phenomena."

A very instructive example of structural stability and singularities we owe to CALLAHAN (1974): Take a piece of fabric at one corner and put it onto a flat surface. What forms can occur locally on the sheet (Fig. 4.1)? There are three possibilities:

a) the sheet lies flatly and smoothly, these points are called the regular ones;

b) a fold line appears on the sheet;

c) a pleat forms at the end of a fold line.

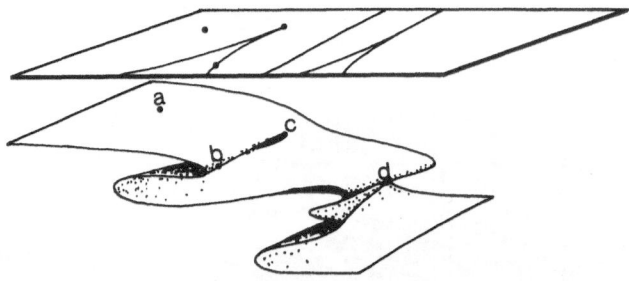

Fig. 4.1: Local forms on a folded sheet (after CALLAHAN, 1974); (a) a regular point, the sheet lies flat; (b) a singular point on a fold line; (c) a singular point where the fold turns into a pleat; (d) a singular point which is not structurally stable.

The points (b) and (c) are called the singular points because of their special nature; in particular, they are structurally stable singular points. To see this, take a point such as (d) in Fig. 4.1. This is an additional type of singular points, but it is not a structurally stable one because any disturbance, slight as it may be, turns this point into one of type (a), (b) or (c). On the other hand, points of types (a), (b) or (c) cannot be made to disappear by a small perturbation. One can dislocate a fold or pleat by a small perturbation, but one cannot affect its presence. The discussed singularities yield in addition a structural information. Put a stiff sheet of paper beside the fabric in the same way: It will only consist of regular and perhaps fold points. It is distinguished from the piece of fabric by the singular points which may appear under this experiment. It is usually the set of stable singularities which allows us to classify objects.

It was THOM (1975) who pushed forward these concepts in mathematics and their applications. He has shown that the concept of structural stability, i.e. the insensitivity to small perturbations, is related in one context to stable singularities. These stable singularities have then been classified by Thom in his "seven elementary catastrophes". Here, examples are given which demonstrate that these concepts are useful to understand patterns and to determine the geometric singularities that an unknown surface generically impresses on a sensing wavefield (DANGELMAYR & GÜTTINGER, 1982). The examples are derived from Huygens' construction of wave fronts and envelopes. A detailed discussion of the mathematical machinery and of mainly physical applications can be found in LU (1976), POSTON & STEWART (1978), GÜTTINGER & EIKEMEIER (eds. 1979), STEWART (1981, 1982).

Fig. 4.1 repeats Fig. 2.1 with the addition of the typical singular points which uniquely determine the local structure of the surface -- i.e. the set of singular points on surfaces provides a skeleton of essential structural information. In the previous chapters such sets of singular points were viewed as instabilities, in this chapter the inverse problem will be studied, the classification of "form" by the intrinsic set of singular points.

In the first section singularities are discussed, which occur as discontinuities in the two-dimensional images of three-dimensional objects. The typical singularities provide useful information in picture processing. The concept of skeletons then relates the analysis of two-dimensional images to the previous discussion of optimization problems on one hand, and to D'Arcy Thompson's classical 'transformations of form' on the other.

In the second section the theory of elementary catastrophes is applied to the linear ray model in reflection seismology. Most of the discussion remains restricted to the two—dimensional track line problem, and the concepts are derived from simple assumptions with basic mathematics. Instead of dealing with waves directly, the linear ray patterns and their caustics are studied. The wave fronts then are analyzed in terms of a continuous plane map. Finally, the traveltime record is analyzed as a map which transforms the three-dimensional spatio-temporal system into the traveltime record. It will turn out that the traveltime record is locally equivalent to (oblique) sections through a swallowtail catastrophe which is located on an oblique line in the depth/time coordinates.

In the third section some aspects of folds and faults are discussed in terms of 'parallel surfaces'. This discussion takes up the 'pre-computer' analysis of large scale deformations in tectonics and reviews this 'early' kinetic approach in terms of recent developments in singularity theory. The lateral continuation and the depth limit of folds will be discussed in different ways and continues in some respect the problems of chapter 2.

4.1 IMAGE RECOGNITION OF THREE-DIMENSIONAL OBJECTS

A common problem in pattern recognition is the reconstruction and classification of three-dimensional objects from two-dimensional pictures. Typically, this problem arises in transmission microscopy. Fig.4.2 shows the larva of a medusa in transmitting light. Locally triangular patterns of increased density appear, which can be related to a bound-

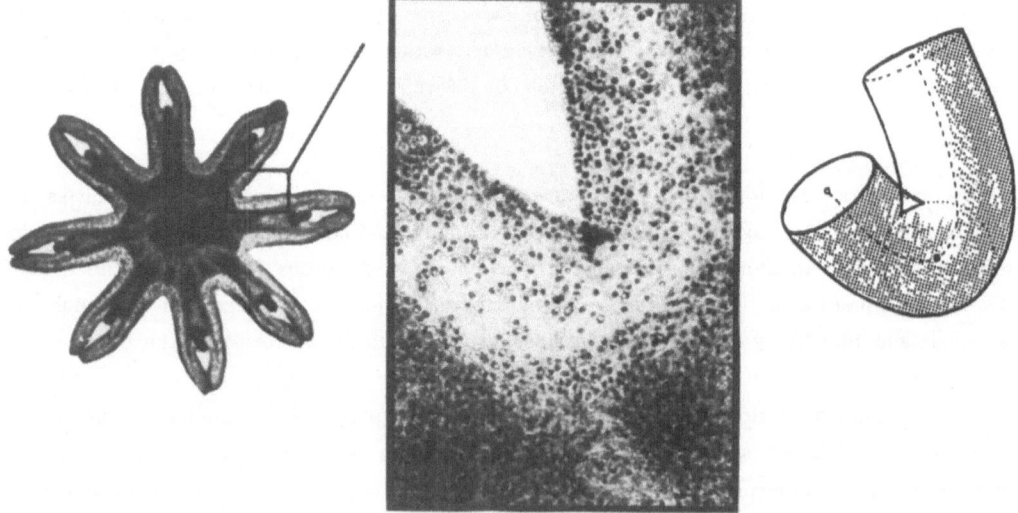

Fig. 4.2: The larva of a medusa in transmitting light. A swallowtail singularity occurs in the two-dimensional image. The identical pattern is found in the two-dimensional image of a transparent canal surface (modified after WUNDERLICH, 1966).

ary effect. An identical pattern is well known from two-dimensional perspective views of locally convex surfaces, especially from canal surfaces in constructional geometry (WUNDERLICH, 1966).

Another field, where similar patterns play some role, are tectonically flattened and deformed images. The proper reconstruction of such images yields useful information for the paleontologist as well as for the determination of the local tectonical stress field. The normal approach in such cases are affine transformations. These describe the deformations sufficiently if the dimension of the object, i.e. of its image, is not altered: if one has maps $R^2 \to R^2$ or $R^3 \to R^3$. In the case of a map $R^3 \to R^2$, new patterns occur, discontinuities at the boundary lines of the image which are closely related to the local surface structure.

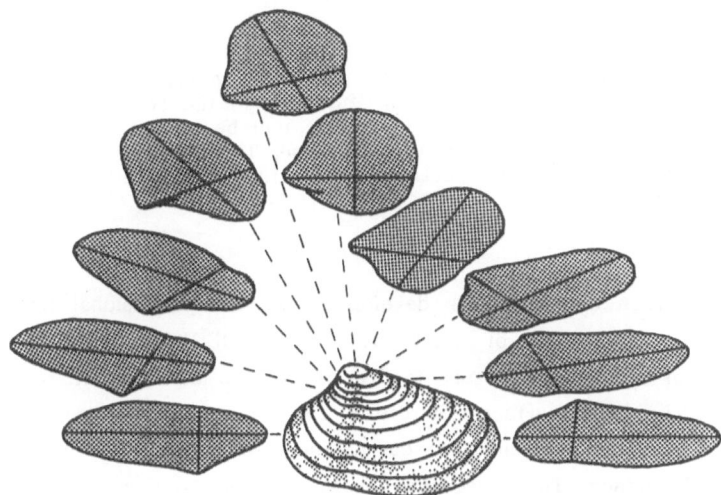

Fig. 4.3: Tectonically deformed and flattened bivalves show surface discontinuities similar to those in Fig. 4.2: In this case, the objects are not transparent, and, therefore, only parts of the surface discontinuities are visible (modified after ROLLIER, 1918).

4.1.1 The Two-Dimensional Image of Three-Dimensional Objects

Locally hyperbolic surfaces are capable to project onto swallowtail-like images in two dimensions (Fig. 4.4). These images are closely related to Thom's swallowtail catastrophe (see e.g. POSTON & STEWART, 1978). This relationship secures that the pattern is structurally stable, and, therefore, it can be used to detect locally concave surface elements or holes in the two-dimensional images of three-dimensional objects.

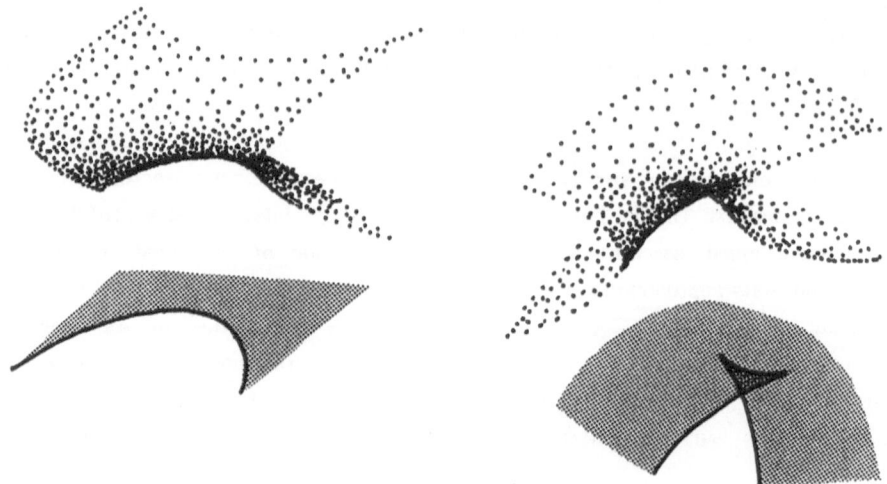

Fig. 4.4: Two projections of a hyperbolic surface element. An oblique projection causes the occurrence of a surface discontinuity in the two-dimensional image space. The three-dimensional surface element can be identified with the catastrophe manifold of the swallowtail catastrophe. The discontinuity is the two--dimensional image of the bifurcation set of the swallowtail (modified after POSTON & STEWART, 1978).

How such surface discontinuities develop in the two-dimensional image space, can be analyzed most simply by use of canal surfaces (WUNDERLICH, 1966). A canal surface is the envelope of a (one-parameter) family of spheres of constant radius (GUGGEN-HEIMER, 1977). The canal surface, therefore, can be generated by moving the center of a sphere along a (three-dimensional) curve; the envelope surface of the spheres then defines the canal surface. The most simple case of such a canal surface is the torus where the leading curve is a circle. Under an oblique projection into the plane the circular leading curve transforms into an ellipse. This deformation of the leading curve can be described by an affine transformation which takes the circle into an ellipse. But the boundary lines of the torus in the two-dimensional projective space behave differently. The generating spheres project onto R^2 as circles independent of a rigid rotation of the torus. The result is that the apparent boundaries of the transparent torus develop into a curve with a self-intersection and with two cuspoid edges, the earlier noticed 'triangle' (Fig. 4.2). Fig. 4.5 gives three different views of the torus and of the local surface discontinuities in the projective plane.

The identical pattern appears if one keeps the leading curve constant and alters the diameter of the canal surface (Fig. 4.6). In this case, classical constructional geometry (WUNDERLICH, 1966) tells us that the locally discontinuous image appears when the circular projections of the spheres enter the evolute of the leading curve, along which the centers of the spheres are located. A family of such surfaces with variable

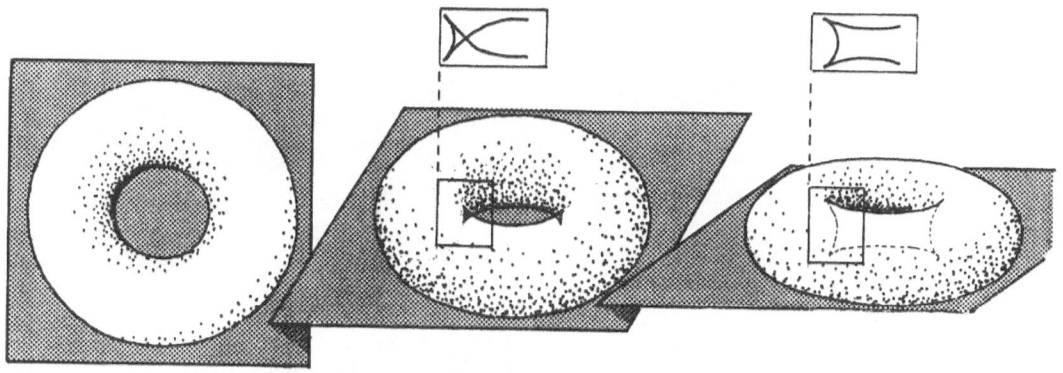

Fig. 4.5: Three perspective views of a (transparent) torus and of its apparent surface discontinuities. The latter develop at the interior hyperbolic surface points, as the projection becomes oblique.

canal diameter maps, therefore, onto a family of involutes or of parallel curves, which unfold inside the evolute of the   leading   curve into swallowtail-like curves (Fig. 4.6).

The latter observations allow to relate the evolution of the image boundaries to Huygens' principle (Fig. 4.7). If the image of the canal surface is studied in two dimensions, then the envelope of the spheres reduces to the envelope of circles with a constant radius. A change of the diameter of the canal surface corresponds to a change of the radii of these circles. In two dimensions, this is identical with a continuous transport of the boundary lines along the normals of the   leading   curve. The image changes locally when the area is reached where the normals intersect (Fig. 4.7). This constructional principle allows to study all possible images of canal surfaces that can occur in two

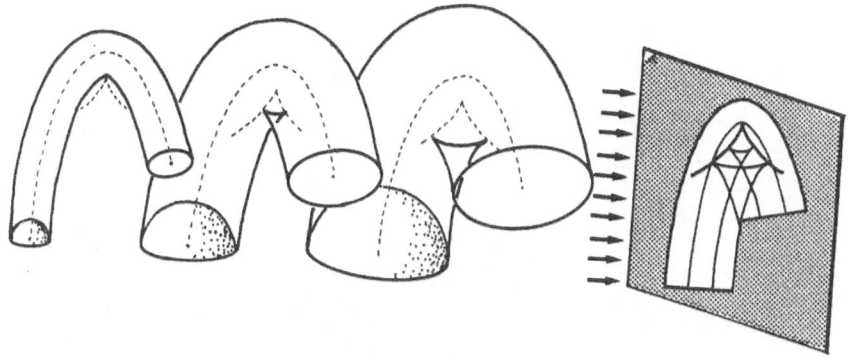

Fig. 4.6: Canal surfaces with identical generating curve but of different diameter. The boundaries of the surface map onto a family of parallel curves in the two-dimensional image space.

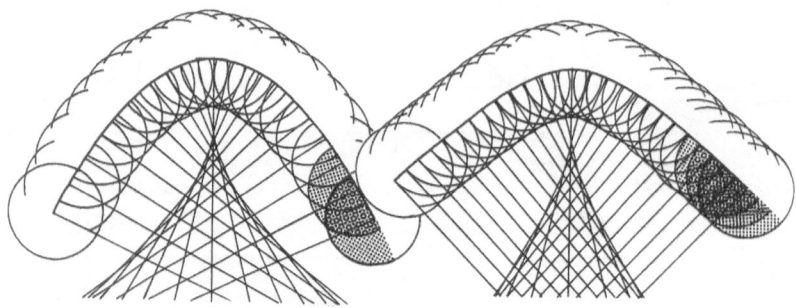

Fig. 4.7: Construction of the two-dimensional images of canal surfaces by use of Huygens' principle. The rays indicate the dislocation lines for the surface boundary in the projective plane.

dimensions (Fig. 4.8). The only discontinuities, which appear, are the swallowtail-like singular curves. The same holds for the perspective views of the torus; the only difference is that, in this case, not the diameter of the canal surface is altered but the local curvature of the leading curve. The result is the same: As the curvature decreases, the surface 'moves into the evolute' of the leading curve, and the boundary lines deform in the identical way discussed above.

The critical points of a canal surface are the hyperbolic points. They behave under the projection onto a two-dimensional image in the discussed way. The occurrence of the 'swallowtail' is typical for locally convex structures, as will be shown in section 4.2. The relation to hyperbolic surface elements allows for a simple construction of

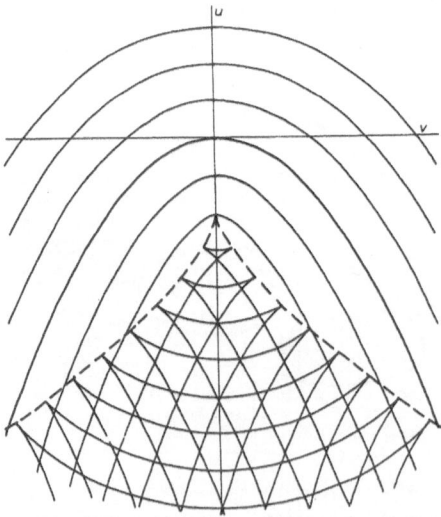

Fig. 4.8: A family of boundary lines for canal surfaces which have a common parabolic generating curve.

151

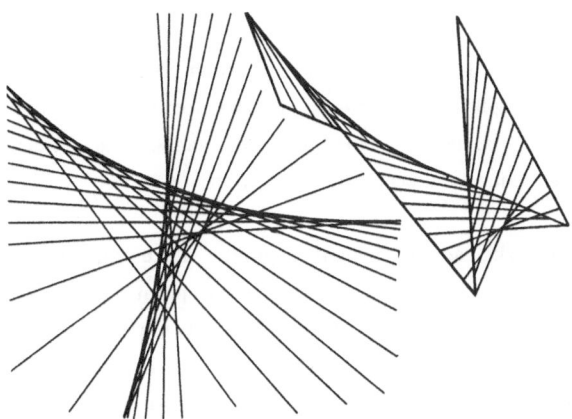

<u>Fig. 4.9:</u> Construction of the swallowtail singularity as a ruled surface over a folded polygon.

the swallowtail. A hyperbolic surface, which locally takes the form z = xy, can be generated as a ruled surface. If these rulings are projected into two dimensions (Fig. 4.9), then the envelope of the straight lines gives again the various possible images of the surface in two dimensions including the swallowtail discontinuity.

The concept of parallel curves, which are generated by a transport along their normals, can be related to the concept of potentials and to elementary catastrophes. In the latter case, the original three-dimensional object resembles the catastrophe manifold. If the leading curve is continuous, it can be locally described as an implicit function f(x,y) = 0. In addition, one can suspect that the family of parallel curves, which is generated by the transport along the normals, can be given in implicit form as c = f(x,y). But this latter formulation can be interpreted as a potential. The directional derivatives, the gradient of this local potential, define the dislocation field. Catastrophe theory then gives a classification of the stable singularities of such 'local' potentials. One of the catastrophes, which appear under the map $R^3 \rightarrow R^2$, is the swallowtail. In the present context it is a stable discontinuity, which allows to identify local hyperbolic surface elements in the two-dimensional image. A more detailed discussion of this singularity will follow in the next sections.

4.1.2 The Skeleton of Plane Figures

In Picture Recognition the problem arises to represent objects economically in the computer, and to classify objects as equivalent even if they are disturbed to some extent. Especially the latter problem leads to the concept of skeletons or 'medial axes' (BLUM, 1973; BLUM & NAGEL, 1977). BOOKSTEIN (1978) gives the following definition of the skeleton of a plane figure:

> The <u>skeleton</u> of a plane figure is a certain graph inside the figure, together with a function on the graph. The skeleton is the locus of all points which do not have a unique nearest boundary point upon the shape; the function is the distance to any of the set of equally distant nearest points.

Thus, the skeleton is just the set of singular points discussed in section 2.4.5, which arose from the problem to find the nearest boundary point. The most natural way of finding the skeleton of a given object is by shrinking it until it reduces to its skeleton (ROSENFELD & WESZKA, 1980). This shrinking can -- theoretically -- be done by a wave front, which starts instantaneously at every point along the boundary and moves with constant velocity -- i.e. by Huygens' principle, as discussed in the previous section (for technical details of computation see e.g. ROSENFELD & WESZKA, 1980). Still more illustrative is the "grassfire" model (BOOKSTEIN, 1978):

> *"We imagine a shape boundary to be 'drawn' on the dry grass of a prairie, and fired (simultaneously). The fire will burn evenly in all directions from its starting locus until it encounters points at which it arrives simultaneously from two directions, whereupon it quenches itself, as grass cannot burn twice. Such loci comprise the skeleton, and the function we seek is the time it takes the fire to arrive there and go out."*

The critical set of singular points we encountered in the optimization problem (section 2.4.5) has now a totally different quality: Together with a vector valued function (which defines the direction to the nearest boundary points) and a distance measurement (which defines the location of the original boundary line) the set of singular points or the 'medial axis' generically defines our object. Fig. 4.10 shows various examples of plane figures and their 'medial axes'.

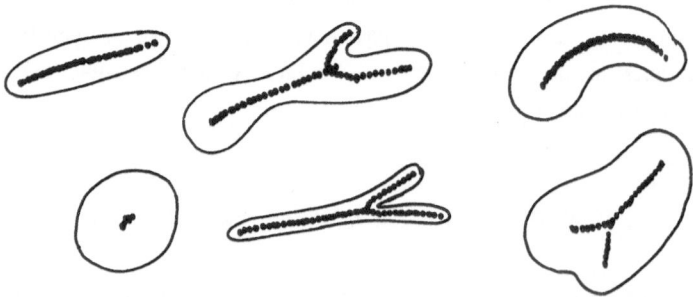

<u>Fig. 4.10:</u> Skeletons (medial axes) of various two-dimensional objects.

4.1.3 Theoretical Morphology of Worm-Like Objects

The concept of skeletons becomes especially simple in the case of cylindrical objects when parts of the boundary line and the skeleton are parallel curves. Fig. 4.11 illustrates the case of a cylinder with spherical caps terminating its ends. We may think about such a cylindrical object in terms of a wiggly worm or a chromosome if we allow

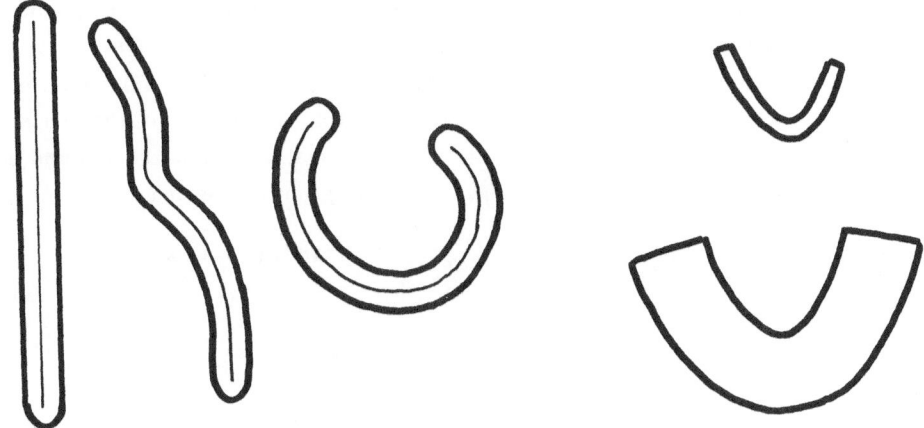

Fig. 4.11: Morphology of a wiggly worm, which preserves length, width, circumference, and area when it bends (a,b). The curvature of coiling, however, is limited by the width of the worm (c).

X- and Y-like patterns. Now, if our wiggly object stretches or changes its form in some way, "our intuition expects that the width of its form will stay practically the same and we expect in addition its length to be invariant" (BOOKSTEIN, 1978). What we then need are precise definitions of width and length and a map which preserves these properties if the 'worm' bends.

Let the cylindrical object be straight, then we always can find a coordinate system so that the boundaries are described by the map

$$x = s \atop y = \lambda \qquad \text{if } |s| < a \tag{4.1a}$$

and

$$x = \pm(a + \cos(s)) \atop y = \lambda\sin(s) \qquad \text{if } a < |s| < a+\lambda . \tag{4.1b}$$

In this case, the medial axis is the line y=0. If the medial line is bended to some arbitrary form, we can describe this deformation by a map

$$x \; ---> \; (F(x),G(x))$$

or
(4.2)

$$u = f(s)$$
$$v = g(s) \; ,$$

whereby we have to secure that the length of the medial line does not change (even locally). This requires that the disturbed image of the medial line is parametricized by arc length, what will be assumed for the following discussion. The next condition to be satisfied is constant width. In the straight object (equation 4.1) width is defined perpendicular to the medial axis. If we transfer this definition to the wiggly worm, then width is measured along the normals to the medial axis, i.e. the deformation (4.2) defines the deformation of the entire object which is given by the map

$$(x,y) \; ---> \; (F(x),G(x)) \pm \lambda \vec{N}(F(x),G(x))$$

or
(4.3)

$$u = f(s) - \lambda \dot{g}(s)$$
$$v = g(s) + \lambda \dot{f}(s) \; ,$$

where N is the normal to the medial line. The spherical caps (4.1b) are still located at the endpoints of the medial line and are not deformed (cf. Fig. 4.11).

Finally, we are interested how the circumference of the object and its area are altered. From equation (4.3) we find the length increment ds of the parallel pieces of the boundary to be

$$ds_B = (1 - \lambda k(s)) \; \sqrt{(\dot{f}(s)^2 + \dot{g}(s)^2)} \; ds \; ,$$
(4.4)

where $k(s)$ is the curvature of the medial line and
$\sqrt{(\dot{f}(s)^2 + \dot{g}(s)^2)} = 1$ because the medial line is parametricized by arc length ,

and the length of the boundary is

$$L_B = \int(1-\lambda k)ds + \int(1 +\lambda k)ds + 2L_{spherical\ caps}$$

$$= 2 \int ds + 2L_{spherical\ caps} \; ; \quad k: \text{curvature of medial axis.}$$
(4.5)

That is, the length of the boundary is simply twice the length of the medial axis, independent of the deformation of the medial line! By definition, the length of the medial line does not change under arbitrary deformations, and the length of the circumference is an invariant property of the map (4.3). The area of the object is given by

$$A = \iint (1-\lambda k)\,ds\,d\lambda + \iint (1+\lambda k)\,ds\,d\lambda + 2A_{\text{spherical caps}}$$
$$= \iint ds\,d\lambda + 2A_{\text{spherical caps}},$$

and again the deformation does not affect the area which, therefore, is another invariant for any fixed value of the parameter λ. The map (4.2) describes ideal deformations -- bending and coiling -- of worm-like objects, which preserve the length of the medial axis, the length (surface) of the boundary, and the area (volume) of the object -- properties which -- by intuition -- can be expected for biological objects.

However, if we return to the discussion of the preceding section, it is clear that there are limits for the deformation of such worm-like objects. As the curvature or the width of the object increases (Fig. 4.6), the surface would develop self-intersections, which cannot be realized. These singularities, which occur at the convex areas of the wiggly worm (Fig. 4.11), are 'standard catastrophes', which will be discussed in more detail in the next sections. For our worm-like object, however, we find a strong correlation between width and the curvature of coiling.

4.1.4 Continuous Transformations of Form

D'Arcy Thompson introduced the concept of shape transformations to describe the morphological evolution in phylogeny and ontogeny (THOMPSON, 1952). BOOKSTEIN (1978) summarizes:

> *"The formal theme of D'Arcy Thompson's method is this: to represent a change of one shape into another by the single mathematical object which is the map of one shape onto the other, and then to visualize this mathematical object."*

This program has been followed with much emphasis, and a common goal was to find the 'ideal family of transformations', which solves the biological problem, as the ideal hydrodynamic problem is solved by the conformal mappings. Many published attempts, of course, relate D'Arcy Thompson's problems to classical physical methods -- conformal mappings (Fig. 4.12; RICHARDS & KAVANAGH, 1947), biorthogonal grids (BOOKSTEIN, 1978), and to the Navier-Stoke equation as a general solution (GRENANDER, 1976).

The methods, based on D'Arcy Thompson's program, compare different objects and morphological states. They describe a map from one state to the other but not a continuous deformation as it is usually obvious in ontogeny and appears likely for

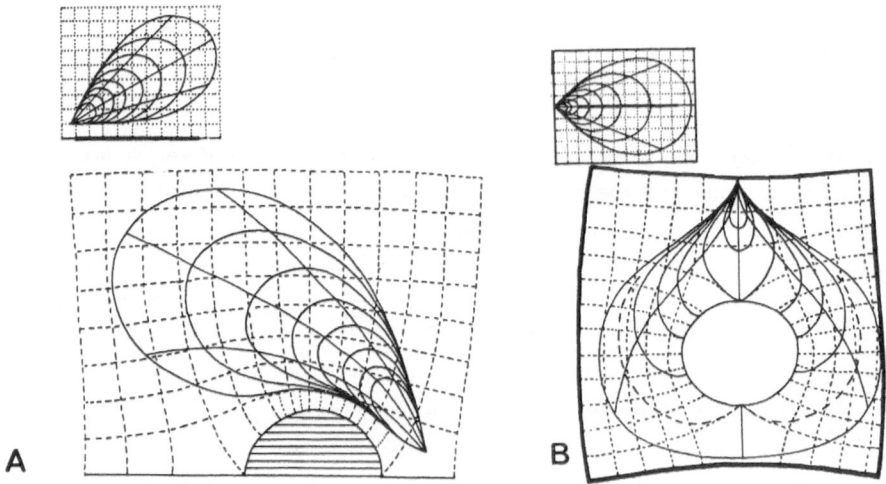

Fig. 4.12: Transformation of form in a bivalve (A) and a brachiopod (B). Both mappings are produced by the conformal map w=a(z + 1/z) + b log z. (Adapted from BAYER, 1978).

phylogeny if one does not believe in macro-mutations. Further, the use of 'Cartesian grids' implies a continuous, a topological deformation. However, in the real systems we find sometimes rather sudden changes although the deformation is continuous.

Equation (4.3) provides a 'prototype' of such continuous deformations, which are capable of sudden morphological changes as soon as the surface enters the 'caustic' of normals. Of course, it is not hard to argue that real biological transformations are much more complicated than the map (4.3). However, if we do not attempt to describe the entire morphology at once but restrict the study to interesting local patterns, then

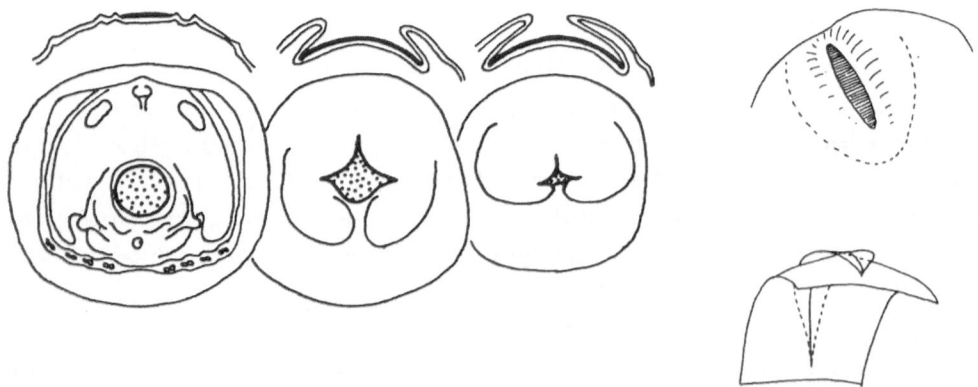

Fig. 4.13: The early ontogeny of *Sepia* (after NAEF, 1928 and BLIND, 1976) and the cicatrix of *Nautilus* (after BLIND, 1976) compared with a swallowtail surface (after THOM, 1970).

157

the map (4.3) provides a first order approximation for the evolution of a locally cylindrical surface element -- what may happen, is the evolution of folds on the surface (cf.
Figs. 4.1--4.9). Fig. 4.13 illustrates folding processes in the early ontogeny of cephalopods,
which can be related to the evolution of a wave front -- the surface -- which folds
and generates singular lines.

The map (4.3) describes a simple case of a wave front and is equivalent to THOM's
(1970,1975) approach to morphogenesis, as soon as not the morphology itself but the
singular sets are emphasized. In the simple example of equations (4.3) the interesting
singularities are the cusp and swallowtail catastrophes. How these elementary catastrophes
are related to the map (4.3), will be discussed in the next section.

4.2 SURFACE INVERSIONS IN THE SEISMIC RECORD --
THE CUSP AND SWALLOWTAIL CATASTROPHES

In a wide field of applications geologists are concerned with the problem to interpret the records of reflection seismology in a qualitative way. The morphological (geometrical) interpretation is here usually much more important than the reconstruction of true
depth relationships -- a major problem for the seismologist. To solve the interpretation
problem quantitatively requires to solve the full dynamics of the wave equation, which
captures the process of remote sensing. In the most general sense, waves are spreading
processes which satisfy a total hyperbolic differential equation (COURANT & HILBERT,
1968). The trouble is that one does not only need the initial conditions but also the

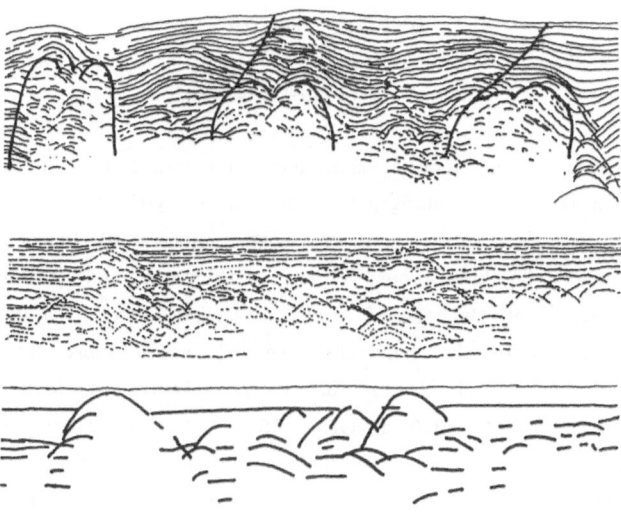

Fig. 4.14: Sketches of traveltime records, which have been interpreted as
salt domes with self-intersecting reflectors (above: after DRIVER & PARDO,
1974; below: after BIJU-DUVAL et al., 1974).

entire boundary conditions to solve the general equation. Instead of solving the total hyperbolic equation one can pick a visible structure from the wave field, say a crest or trough line or, in a general notation, a wave front (WRIGHT, 1979). To solve the spreading process of the wave front one can use Huygens' principle (COURANT & HILBERT, 1968; OFFICER, 1974). If a wave front is known at a certain time, one studies the evolution of the wavelets that spread from every point of the generating wave front. The successive wave fronts are the envelopes of the wavelets. In the three-dimensional case the wavelets are spheres, and the successive wave fronts are surfaces $F(x,y,z)$ = constant. If the propagation of the wavelets is fairly constant along the generating wave front, the successive wave fronts can be approximated by a transport of the original surface along its normals with constant velocity. In two dimensions the spherical wavelets reduce to circular ones, and the problem is moderately simplified. In a mathematical sense this construction of the propagating wave fronts can be described as a continuous map, and, at least locally, one can suspect that this map can be summarized in a potential equation $V = w(x,y,z)$.

The interesting structures of these potentials are their stable singularities. The geometrical singularities can well be studied by topological methods. DANGELMAYER & GÜTTINGER (1982) discussed the physical aspects of remote sensing -- Fresnel-zone topographies, diffraction patterns etc. -- carefully in terms of catastrophe theory. They showed that the topological approach yields reasonable results for the inverse scattering problem as well as for the on-site survey:

> *"Since, in practice, the analytic analysis is of little interest to the seismologist -- because he needs an overall picture, a 'Gestalt' point of view to classify his forms -- tackling the inverse problem at its topological roots comes much closer to the interpreter's intuitive geometric, i.e. qualitative, approach"* (DANGELMAYER & GÜTTINGER, 1982).

This "qualitative approach" is, indeed, still more important for the geologist than for the seismologist. Interpreting seismograms from areas with salt domes it may be a reasonable question whether intersecting reflectors are realistic or singular patterns within the seismogram. Shadow zones, double reflections and 'hyperbolic reflections' are well known patterns of the record. Figs. 4.14 and 4.15 give some examples how such 'anomalies' show up in the traveltime record. Much more impressive structures of these types have been described from 3.5 to 12 kHz records (echograms) (JOHNSON & DAMUTH, 1979; DAMUTH, 1980;FLOOD, 1980; EMBLEY, 1980). The interpretation of echograms and of patterns like 'hyperbolic reflections' became of interest during the last decade, when sedimentologists started to interpret the deep sea morphology for their purposes.

In the case of high frequency echograms, the wave front evolution can be well

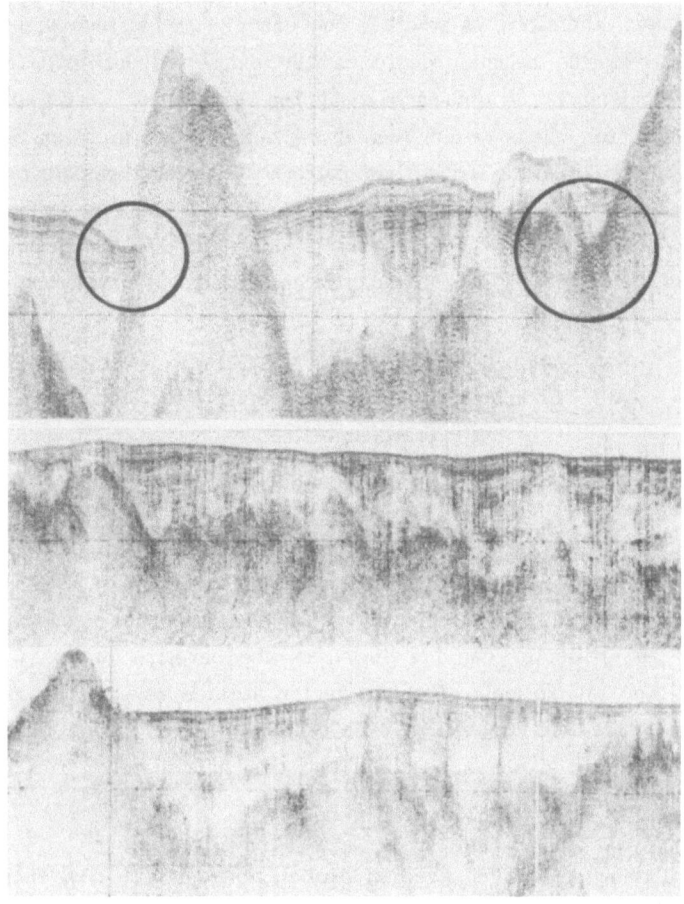

Fig. 4.15: Traveltime records with intersecting reflections, 'extensions of sedimentary layers into the basements' and high energy zones within an isotropic sediment cover.

approximated by linear rays (FLOOD, 1980). In the case that the model is reduced to the 'seismic track', the problem is further simplified. The relation between a (cylindrical) surface and the traveltime record along the track line can be viewed as a map $R^2 \longrightarrow R^2$ or $(x,y) \xrightarrow{s} (u,2t)$, where (x,y) are the spatial coordinates of a section through the surface and $(u,2t)$ are the space-traveltime coordinates of the record. In this notation, the interpretation of the seismogram becomes identical with the problem to determine the map 's'. This approach was recently stressed by FLOOD (1980), who analyzed periodic wavefields. He found that 'hyperbolic reflections' depend on the wavelength of sinusoidal surfaces and water depth. But the approach by any global surface approximation is much too general: There is no real chance to establish a sufficient catalogue of global morphologies. One problem, therefore, is whether one can classify surface points in such a way that their image on the seismic record can be uniquely identified. This, of course, is a topological problem. It is the purpose of this section to derive a classification of traveltime images for the two-dimensional (seismic track) problem. This classification will be one which relates the traveltime patterns to local topological properties of the reflector. There are two ways to discuss the relation between surface properties and the traveltime image, versus the wave fronts or versus the plane map approach. Both methods will be discussed to demonstrate different aspects of catastrophe theory.

4.2.1 Computer Simulations of Rays, Wave Fronts and Traveltime Records

Ray theory becomes drastically simplified if one assumes that the rays are straight lines, and that the source and the receiver are located in a single point. This situation is nearly realized for deep sea echograms (FLOOD, 1980), but it can also be used as a first approximation for other reflection seismograms. Under these special conditions, the surface of the reflector can be viewed as the envelope of the wavelets, it forms a 'wave front', and the normals of the reflector are the rays along which the reflected wave front propagates (e.g. GRANT & WEST, 1965). If the section through the reflector along the track line is given analytically, one can immediately write down the (linear) ray equations.

If the reflecting surface is given in explicit form as $y = f(x)$, then the <u>linear rays</u> are given in parametricized form as

$$u = x_0 - \tau f'(x_0)$$
$$w = f(x_0) + \tau$$

$$(4.6)$$

where $(x_0, f(x_0))$ defines a point on the reflector, (u,w) are the spatial coordinates of the ray which passes through the point $(x_0, f(x_0))$, and τ is a parameter which generates the ray. It turns out that the patterns, which can be found, are not a property of the rays, but that they need to be formed by the morphology of the reflector (WRIGHT, 1979). The reflector impresses its pattern generically onto the sensing wave system (DANGELMAYER & GÜTTINGER, 1982), in this case, onto the family of rays.

Equation (4.6) allows to draw the linear ray pattern for any differentiable function. This was done for a sinusoidal function in Fig. 4.16, and it becomes clear that the seismic record will be disturbed in the areas of ray overlap. In these areas one finds double and triple reflections. The two pictures of Fig. 4.16 show the farfield pattern and, en-

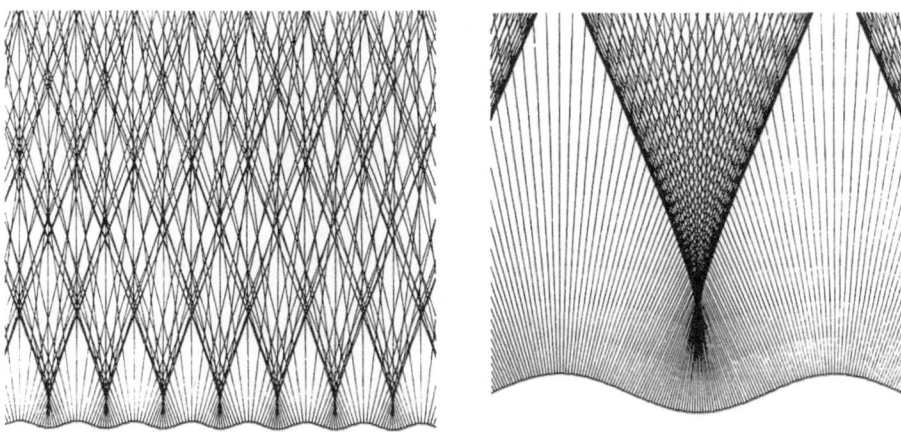

Fig. 4.16: Computer simulations of the linear ray system of a sinusoidal reflector. Left: farfield pattern, right: nearfield pattern.

larged, the nearfield pattern (close to the surface). Fig. 4.17 illustrates that the rather complicated farfield pattern results from superpositions of various nearfield patterns.

If the rays, which pass through the surface element of a wave front, are known, then the evolution of the wave front along the ray family can be simulated, after the ray equations have been normalized by arc length. In the case of linear rays one finds the successive wave fronts from the continuous map

$$u = x - \tau f'(x)/\sqrt{(1+f'(x)^2)}$$

$$w = f(x) + \tau/\sqrt{(1+f'(x)^2)}.$$

(4.7)

(u,w) are the new spatial coordinates of a point $(x,f(x))$ on the original wave front, and τ is the distance between these two points ($\tau = vt$, v: velocity, t: time). The equations (4.7)) define a continuous map from the plane into the plane

$$\{(u,v) \in (x,y) \mid \xrightarrow{S(\tau)} (u,v)\},$$

(4.8)

where (x_o,y_o) are the points on the (cylindrical) reflector. Fig. 4.17 illustrates how the wave fronts evolve from sinusoidal surfaces, and how they are folded within the areas in which the rays intersect. As was noted earlier, the reflector impresses its structure onto the family of rays and onto the wave fronts. This is illustrated in Fig. 4.18 where the reflector was simulated by a cycloid. When the generalized cycloid develops a cusp, a local singular point, then the ray and the wave front pattern change dramatically. Now, it is not hard to see that this is not a structurally stable situation. First one

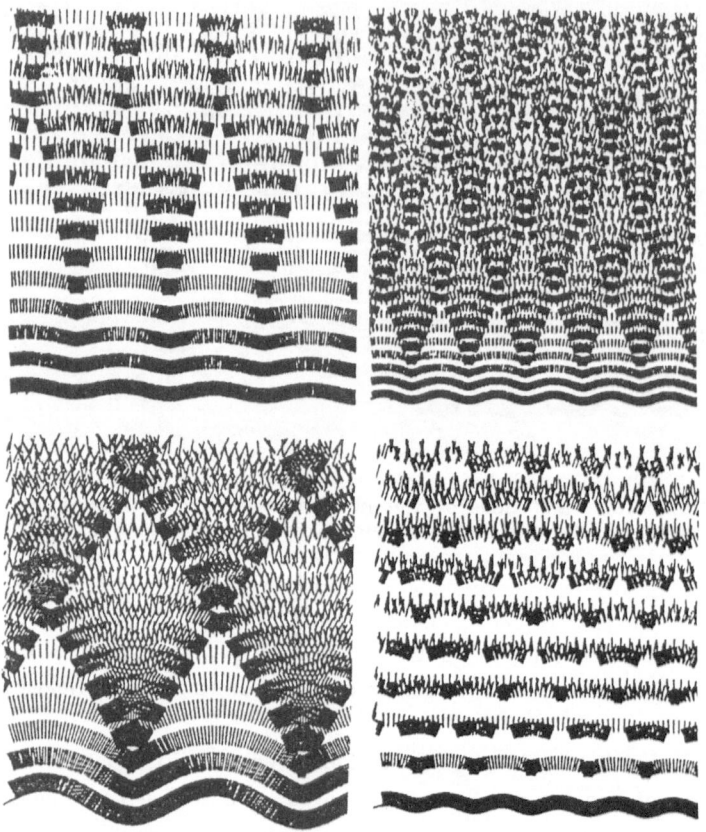

Fig. 4.17: Computer simulations of the evolution of rays and wave fronts from sinusoidal reflectors.

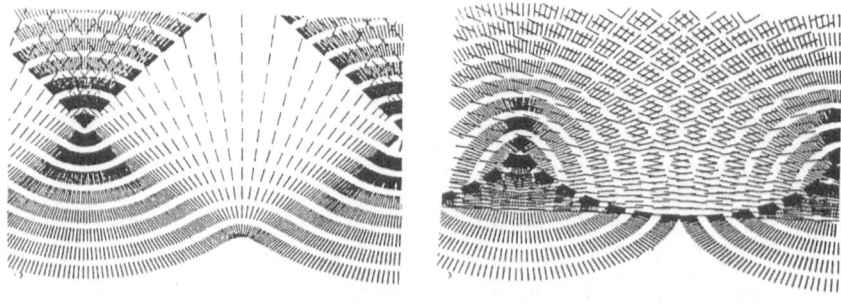

Fig. 4.18: Linear ray system and wave fronts of a cycloid. The degenerated case occurs when the cycloid develops a cusp.

can argue that the cusp is morphologically not stable. Any small disturbance turns it into a pattern like Fig. 4.17. In addition, in this degenerated case, the ray approach does not fit Huygens' principle. Fig. 4.19 illustrates how the envelope of the wavelets evolves near the singular surface point. It turns out that the successive wave fronts 'ignore' the singularity of the reflector. They are again continuous functions, which can be approximated by a surface model like the generalized cycloid of Fig. 4.18. This observation allows to ignore such singularities for most of the following discussion.

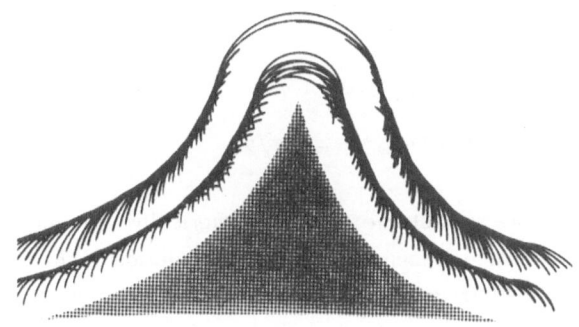

Fig. 4.19: Huygens' wave front construction by wavelets near a singular surface point.

The last point to be discussed is how the surface maps onto the traveltime record. To find the map $(x_o, y_o) \longrightarrow (u, vt/2)$ one has to section the ray pattern in a certain distance of the reflector, i.e. along the survey track line. The modified horizontal coordinate 'u' (Fig. 4.20) is a function of the inclination of the rays. If the track line is taken as the zero level, y = 0, then the horizontal dislocation is

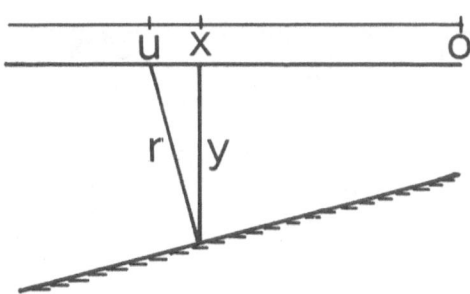

Fig. 4.20: The parameter system for the linear ray model: (x,y,): surface coordinates of the reflector, u: horizontal coordinate of source and receiver at y = 0, r: distance between source (receiver) and surface.

$$\Delta x = (x - u) = -f'(x)f(x). \qquad (4.9)$$

The horizontal dislocations can be easily derived from equations (4.6)) and the condition $y = w = 0$ if the parameter τ is eliminated. Now, the traveltime is proportional to the distance between the shot point and the reflector point (Fig. 4.20) or proportional to

$$r = \sqrt{((u-x)^2 + f(x)^2)}. \qquad (4.10)$$

Thus, if all variables are expressed in terms of x and f(x), one finds the map (FLOOD, 1980) $(x,f(x)) \longrightarrow (u,w)$ as

$$u = x - f'(x)f(x)$$
$$r = f(x)\sqrt{(1-f'(x)^2)} = vt/2. \qquad (4.11)$$

The equations allow to simulate traveltime record numerically for any differentiable surface trace. Fig. 4.21 gives some examples of such simulations. The figures include the ray systems and the simulated traveltime records. The surface models are all convex, and this gives 'hyperbolic reflections' on the traveltime record. The mapping equations, which have been discussed so far, allow to simulate the linear ray pattern, the evolution of wave fronts and the traveltime record for any differentiable surface element. Indeed,

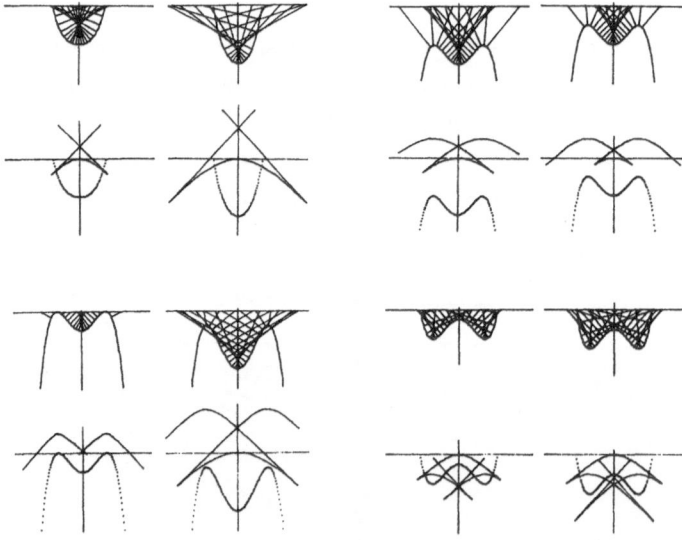

Fig. 4.21: Computer simulations of rays and of the traveltime record for various surface models (fourth order polynomials). The saddle points of the traveltime record have been set to zero depth.

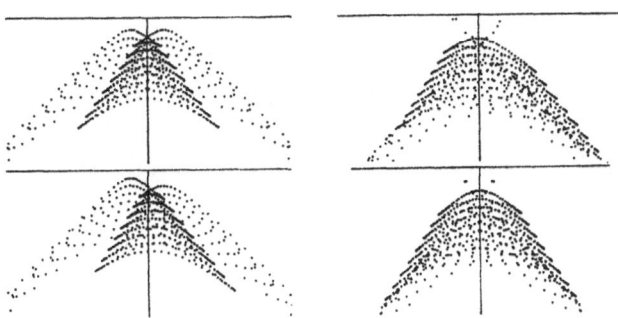

<u>Fig. 4.22:</u> Simulations of the traveltime record like in Fig. 4.21 but as point plots for 'layered media'.

these approximations can be extended to three-dimensional surface structures. But they are still much too general. There is an infinite number of possible functions, which can be used to approximate a reflector surface, and these functions may depend on a large number of parameters so that we are not able to catalogue the traveltime patterns within the parameter space. The major aim of the next section is, therefore, to analyze the local properties of reflectors, and to show how these local properties impress their structure generically on a sensing wavefield and, therefore, on the traveltime record.

The computer simulation has the advantage that more complicated systems can be simulated. Thus, one can transform the geometrical simulation into a point plot (Fig. 4.22) which resembles the received energies to some extent, i.e. the observed traveltime record. The comparison of such a plot for a multilayer-system (Fig. 4.22) with traveltime records (e.g. Fig. 4.15) shows that not only 'hyperbolic reflectors' may arise from local concave surface elements. High energy zones, which are sometimes found in otherwise nearly isotropic areas, may well be related to the surface morphology rather than to a property of the sediment cover.

4.2.2 Local Surface Approximation

In the preceding discussion it became clear that the reflecting surface impresses its structure onto the rays, onto the wave fronts and, therefore, finally onto the traveltime record. It is this generic situation which makes remote sensing a structurally stable process -- and structural stability alone secures that one has a chance to reconstruct the surface properties. On the other hand, it turned out that it is unreasonable to work with global surface structures. Therefore, at first one needs a classification of the critical points on the reflecting surface. This classification was done by DANGELMAYER & GÜTTINGER (1980, 1982) in detail for three-dimensional problems. Here the discussion will be restricted to plane curves, i.e. to a first approximation of cylindrical reflectors along the track line. In this case, the usual situation will be that the function, which describes the surface, is of bounded variation (GUGGENHEIMER, 1977). For most real surfaces we can even suppose that a surface line is of small variation, and, therefore, the usual situation will be that the curve can be locally approximated by a Taylor series

$$f(x) = a_0 + a_1 x + a_2 x^2 + \ldots + \text{higher terms.}$$
(4.12)

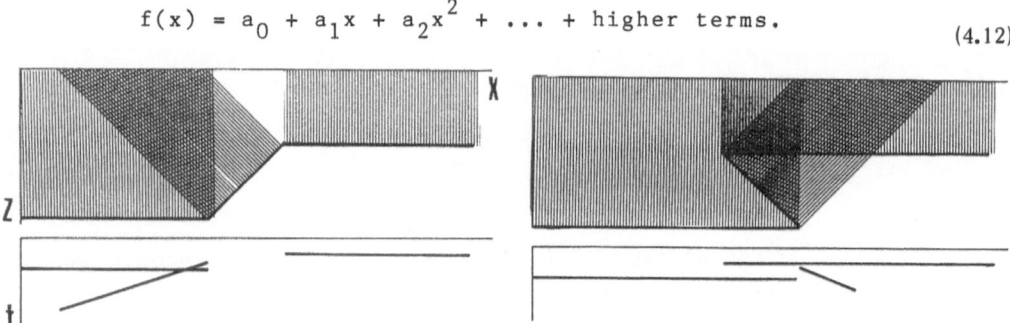

Fig. 4.23: Ray systems (x,z) and 'traveltime record' (x,t) of linear surface elements. The straight lines map onto straight lines, but the horizontal dislocation can cause onlapping features and shadow zones.

Now, there are two interesting cases: If the coefficients $a_2 = 0$ and $a_1 \neq 0$, the reflector equals locally (near $x = 0$) a straight line. This straight reflector element maps onto a straight line on the traveltime record (Fig. 4.23), as can be proved by use of equation (4.11). The only spectacular patterns are summarized in Fig. 4.23. An inclined straight line is dislocated along the horizontal coordinate. This horizontal dislocation can cause shadow zones and onlapping patterns in the traveltime record. Examples for this disturbance of the record are the intersections between sediment cover and basement in Fig. 4.15. On the other hand, the geometrical property 'to be a straight line' is preserved under the map, i.e. there is no spectacular deformation on the traveltime record.

Things become more interesting if the parameter a_2 is non-zero. In this case, the Taylor expansion can be simplified in the following way: One locates a new coordinate system at $x = a_0$ by use of the map $x \longrightarrow x - a_0$. Then one rotates the new coordinate system in such a way that the transformed x-axis coincides with the tangent at $x = 0$, and that the y-axis coincides with the non-oriented normal at this point. The transformation reduces the Taylor series to the form

$$f(x) = \frac{1}{2} k x^2 + \ldots + \text{higher terms.}$$
(4.13)

The parameter k is the local curvature of the curve at the point $x = 0$

$$k = f''(0)/(1-f'(0)^2)^{3/2} \, . \tag{4.14}$$

Dependent on the sign of k the point is either a maximum or a minimum in the local coordinates. The discussed transformation relates the local structure of the reflector to its curvature at the critical point, a well known procedure from differential geometry (GUGGENHEIMER, 1977; DO CARMO, 1976). If the Taylor series starts with higher terms than x^2, one has a degenerated situation, and one will find patterns like in the case of the cycloid of Fig. 4.18. Such situations will be avoided during most of the following discussion.

In the case $0 < k < \infty$ the only spectacular points are local minima, i.e. a locally concave reflector. Under these constraints there exists an area where the rays intersect, and one will receive reflections from several surface points (Fig. 4.16). Therefore, the stable spatial patterns of rays will be analyzed in the next section.

4.2.3 Linear Rays, Caustics and the Cusp Catastrophe

The interesting structures in ray geometry are caustics (e.g. BEN-MENAHEM & SINGH, 1981) -- the envelopes of the rays or the boundary line of the area where rays overlap (Fig. 4.16). In geometrical optics a caustic appears as a line of high intensity (e.g. NYE, 1979), in reflection seismics it separates those areas, which are covered by a single family of rays, from those areas with two or more intersecting ray systems. Within the linear ray model the caustic is found as the locus of the radii of curvature of the reflector line. If the reflecting curve is given explicitly, one has the classical formulae (e.g. GUGGENHEIMER, 1977)

$$\begin{aligned} u &= x - f'(x)(1+f'(x)^2)/f''(x) \\ w &= f(x) + (1+f'(x)^2)/f''(x) \end{aligned} \tag{4.15}$$

or by use of the curvature $k = 1/R$

$$\begin{aligned} u &= x - Rf'(x)/(1+f'(x)^2) \\ w &= f(x) + R/(1+f'(x)^2) \, . \end{aligned} \tag{4.16}$$

The second set of equations relates the caustics to the continuous map (4.7) for the wave front evolution. From the relation $\tau(t) = R(x)$ one finds the time t, at which a certain point of the wave front arrives at the caustic.

Now, if the reflector line is locally approximated by a parabola $y = bx^2$, the equations for the caustic take the form

$$u = 4b^2 x^3$$
$$w = 3bx^2 + \frac{1}{2b}$$

(4.17)

or implicitly

$$27u^2 = 16b(w - \frac{1}{2b})^3.$$

This is a semicubic parabola hanging over the generating convex reflector line. Its only parameter is the local curvature of the reflector (b = k/2). Therefore, the caustic is a stable spatial pattern which is uniquely determined by the local property of the reflector.

The semicubic parabola, which appears here as the caustic, is well known as the critical set of THOM's (1975) cusp catastrophe. Indeed, for the ray patterns discussed earlier this curve bounds the area where the intersection of rays causes abnormal reflection patterns. To see how catastrophe theory is involved we change the viewpoint slightly. In the seismic record one picks up the reflections along a line which is located in a certain height above the reflector. The horizontal dislocation of a reflector point on the record can be found from the equations of the normals (4.6) by elimination of the parameter τ , and one finds the new horizontal coordinate on a track line of height w to be

$$u = x - f'(x)(w - f(x)).$$

(4.18)

By setting w = constant (the track line) one has a relation between the startpoint of the rays and the point where they hit the track line. Now, one can insert the equation for a local parabolic reflector approximation, $y = bx^2$, and from (4.18) one finds the point where the ray intersects the track line:

$$u = x - 2bx(w - bx^2)$$

or

$$u = (1 - 2bw)x - 2b^2 x^3.$$

(4.19)

Because the curvature should not vanish at the spectacular point, one can standardize this equation:

$$\frac{u}{2b} = \frac{(1 - 2bw)}{2b^2}x - x^3$$

or

$$U = x^3 - sx$$

(4.20)

with obvious parameter identifications.

The new parameter 's' is a composite structure of the local curvature of the reflector (b) and of the height of the track line (w) above the spectacular point; thus, a variation of 's' corresponds either to a change of the height of the track line or/and to a change of the local curvature. The resulting cubic equation (4.20) can be simply analyzed. It allows for three possible situations. If s > 0, then the cubic is monotonously increasing (or decreasing), and the map x → u is one to one, i.e. there exists only one family of non-intersecting rays. If s < 0, then the cubic has a maximum and a minimum, and there exists an area where the map x → u is not uniquely determined, i.e. the rays intersect. The case s = 0 defines the transition state between these two possibilities, it defines the loci where the caustic intersects the track line.

Now, the critical area in the (x,s)-space is given by the extrema of the cubic equation. These extrema define the loci on a track line, w = constant, where the caustic intersects the track line. The extrema are given by

$$U_x = 0 = s + 3x^2$$

or

$$s = -3x^2.$$

(4.21)

From equation (4.20) and (4.21) one finds the critical line in the (u,s)-space by elimination of the variable x:

$$u^2/4 = -s^3/27$$

(4.22)

i.e. up to a proper parameter setting the same equation as before. If one relates a change of the parameter 's' to a change of track height (b = constant), then the semi-cubic parabola describes just the earlier discussed caustic. On the other hand, we have already seen that the meaning now is much more general because the parameter 's' includes also changes of the local curvature of the reflector line. Thus, equation (4.20) captures the possible spatial patterns in their most general sense by a minimal set of parameters. One can use equation (4.20) to draw a picture of the critical surface in the three-dimensional space (u,x,s). This folded surface is shown in Fig. 4.24. Every section s = constant through this folded surface describes the dislocation of the horizontal reflector coordinate along a possible track line (for a fixed local curvature).

To arrive at the final catastrophe representation, the viewpoint has to be changed once more. The cubic equation (4.20) can be discussed in terms of its discriminant or in terms of the number of its roots. This gives the additional information how many rays may intersect at a point in the spatial coordinates. The discriminant takes the form

$$D = u^2/4 - s^3/27.$$

(4.23)

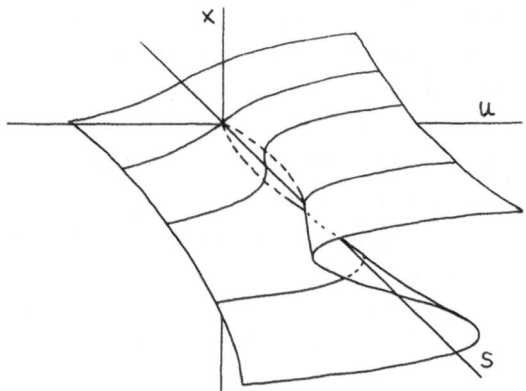

<u>Fig. 4.24:</u> The catastrophe set of the cusp catastrophe.

For D = 0 one has the points which separate the parameter settings leading to a single root (D > 0) from those that cause triple roots (D < 0). Now, the question how many roots a cubic equation has is identical with the question how many extrema a quartic equation may have. The cubic can, therefore, be embedded into the catastrophe potential

$$V = x^4/4 + ux^2/2 + sx, \qquad (4.24)$$

which has been published as the cusp catastrophe (Rieman-Hugoniot catastrophe in THOM, 1975). This catastrophe potential captures the discussed two-dimensional ray patterns in their most general topological behavior.

The previous discussion of the cusp catastrophe allows to classify the traveltime record in terms of depth and local curvature of the reflector line. To do this explicitly one can write the local parabolic approximation as $y = -a + bx^2$, where the parameter 'a' indicates depth. The track line is then located at depth zero. From the equations (4.11) one finds the local traveltime image:

$$u = (1-2ab)x + 2b^2x^3$$
$$r = (-a+bx^2)(1-4b^2x^2)^{1/2}. \qquad (4.25)$$

As turned out from the analysis of the cusp catastrophe, the critical set is given by

$$s = (1-2ab)/(2b^2) = 0. \qquad (4.26)$$

The parameter identification a = -w relates this representation to equation (4.20). Now the critical set can be rewritten in terms of the parameters (a,b) as

171

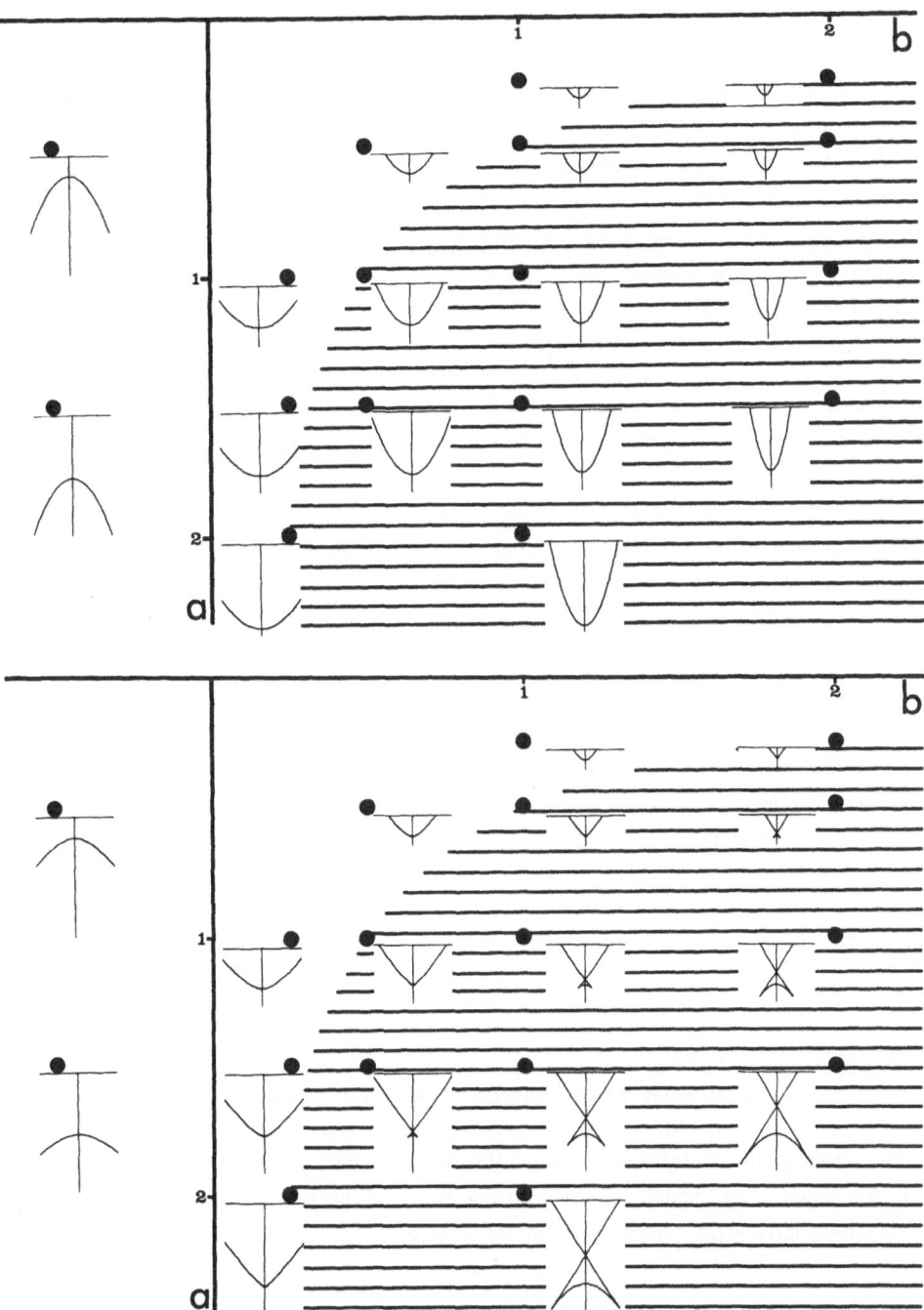

Fig. 4.25: The morphology of local elements on the reflector line (upper graph) and their image on the traveltime record. The surface elements and their images are located in the parameter system depth (a) and local curvature of the reflector line (b). Inside the hyperbolic boundary the reflector image is inverted, i.e. it is the domain of hyperbolic reflections.

$$1 - 2ab = 0. \tag{4.27}$$

Equation (4.27) describes a hyperbolic boundary line in the (a,b)-space, and equation (4.25) allows to compute the image of various parabolas dependent on the choice of the parameters 'a' and 'b'. Fig. 4.25 illustrates the relationship between the local morphology of the reflector and its image on the traveltime record within the parameter space (a,b). From the discussion of the cusp catastrophe we know that the parameter 's' affects the intersection of the caustic with the track line. In analogy to the parameter 's' one can vary equation (4.27):

$$c - 2ab = 0. \tag{4.28}$$

This defines a family of hyperbolae in the (a,b)-space of identical traveltime image (different depth location), which result from different conditions (Fig. 4.25).

Thus, the previous discussion provides us with some practical results, at least for the interpretation of echograms. The analysis of the traveltime record, in terms of local properties of the reflector line and of caustics, allows to classify the traveltime images by a minimal set of parameters, and it becomes clear that these parameters -- depth and local curvature -- are not independent with respect to the effects they produce on the traveltime record. In addition, it becomes clear that the extrema of a reflecting surface are stable points. In this case, one can approximate the reflector line locally by a parabola without any rotation of the local coordinate system, and the point $(0,f(x))$ is recorded at its correctly horizontal position as well as with the correct traveltime. This allows to estimate the wavelength of sand waves and similar structures along the track line from the original traveltime record. Furthermore, the amplitude can be estimated as well in terms of traveltime, and the relation to the cusp catastrophe allows to draw charts, from which the local curvature can be estimated.

4.2.4 Wave Fronts and the Swallowtail Catastrophe

The next point of interest is how the wave fronts evolve near a cuspoid caustic. Within the linear ray model, a wave front is given as a set of points (on the family of rays) which have equal distance from the reflector

$$\{(u,w) \in (x,y) \mid ((x-u)^2 + (y-w)^2)^{1/2} = r = \text{const.}\}. \tag{4.29}$$

This two-dimensional relationship can be extended to the three-dimensional case (DANGELMAYR & GÜTTINGER, 1982). Here it is more appropriate to return to the continuous map for the wave fronts, as it was derived in equation (4.7). From these equations one finds a wave front by setting τ = constant. The only critical set for the rays is the cuspoid caustic. Therefore, as WRIGHT (1979) points out, we should "expect the critical value graph of the cusp catastrophe, which is the bifurcation set of the swallowtail". Indeed, if one draws the wave fronts for several values of the parameter τ to simulate their evolution from a locally parabolic reflector, then they take the form of sections through the swallowtail catastrophe (Fig. 4.26) as far as they are located inside the caustic. To see in detail how the swallowtail is related to wave fronts one

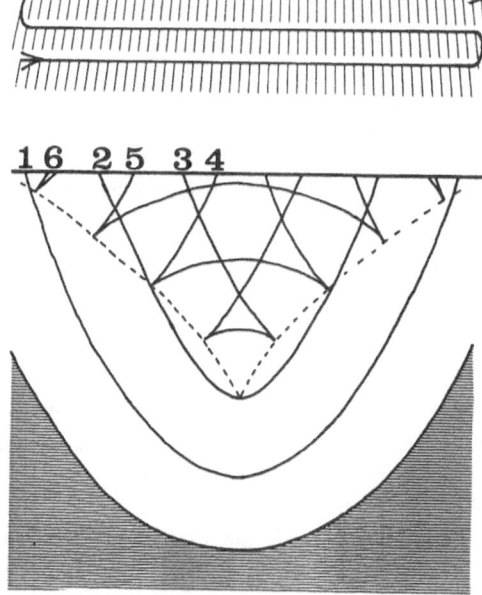

Fig. 4.26: Evolution of parabolic wave fronts into swallowtails. The unfolding of the wave fronts is caused by the folded ray system, which is due to the cusp catastrophe. The numbers indicate values for the parameter of evolution -
= vt.

can develop the equations (4.7) in Taylor series. If the local parabolic surface approximation formula is inserted into equations (4.7), then the Taylor expansion of these equations up to order 4 gives the approximations

$$u = (1-2b\tau)x + 4b^3\tau x^3$$

$$w = \tau + b(1-2b\tau)x^2 + 6b^4x^4. \tag{4.30}$$

If $b \neq 0$ and $\tau \neq 0$, then these equations can be standardized to the form

$$W = sx^2 + 3x^4$$

$$U = 2sx + 4x^3 \tag{4.31}$$

where $\qquad W = (w-\tau)/(2b^4\tau), \qquad U = u/(b^3\tau),$

$$s = (1-2b\tau)/(b^3\tau).$$

On the other hand, the catastrophe potential of the swallowtail is defined as

$$V = x^5/5 + ax^3/3 + bx^2/2 + cx. \tag{4.32}$$

Its critical value graph in the parameter space (a,b,c) is defined by its derivatives (THOM, 1975)

$$
\begin{aligned}
V_x &= x^4 + ax^2 + bx + c = 0 \\
V_{xx} &= 4x^3 + 2ax + b \quad\ = 0 \\
V_{xxx} &= 12x^2 + 2a \qquad\quad = 0.
\end{aligned}
\tag{4.33}
$$

If one uses the first two derivatives to solve for the parameters b and c in terms of a and x, one finds the map

$$
\begin{aligned}
b &= -4x^3 - 2ax \\
c &= \ 3x^4 + \ ax.
\end{aligned}
\tag{4.34}
$$

By a proper choice of the signs (take $x \rightarrow -x$) this map becomes equivalent to the local Taylor expansion (4.31) if one takes the following parameter identifications

$$U \equiv c, \quad W \equiv b, \quad a \equiv s.$$

At least locally (by an approximation up to order 4), one finds that the wave fronts are equivalent to sections a = s = constant through the catastrophe set of the swallowtail. Again, one finds that a standard catastrophe on THOM's (1975) list gives a good approximation to the ray model. In this case, the swallowtail catastrophe describes rather pretty the evolution of wave fronts.

4.2.5   Wave Front Evolution and the Traveltime Record

An examination of the original Taylor approximations for the wave fronts (equations (4.30)) shows that the sections s = const. through the swallowtail are located on a line $w = \tau$. The projections onto the (u,w)-plane ($(u,w) \mathfrak{E}(x,y)$) of these sections through the modified swallowtail (4.30) give the typical evolution pattern for the wave fronts (Fig. 4.26). The sections through the swallowtail are sitting one behind the other in the caustic. Now, the parameter τ can be written as $\tau = vt$ (v: velocity, t: time), and we find that the way, in which the swallowtail is sitting above the cuspoid caustic, depends on the sonic velocity of the medium or on the velocity of the wave front dislocation. On the other hand, the velocity cannot affect the spatial pattern -- the caustic -- as turned out during the discussion of the cusp catastrophe. The caustic is a structurally stable spatial pattern, which only depends on the local surface structure, i.e. the local curvature.

Now, the analysis of wave fronts adds a third dimension, time, to the two-dimensional spatial coordinates. The map from the reflector to the traveltime record, there-

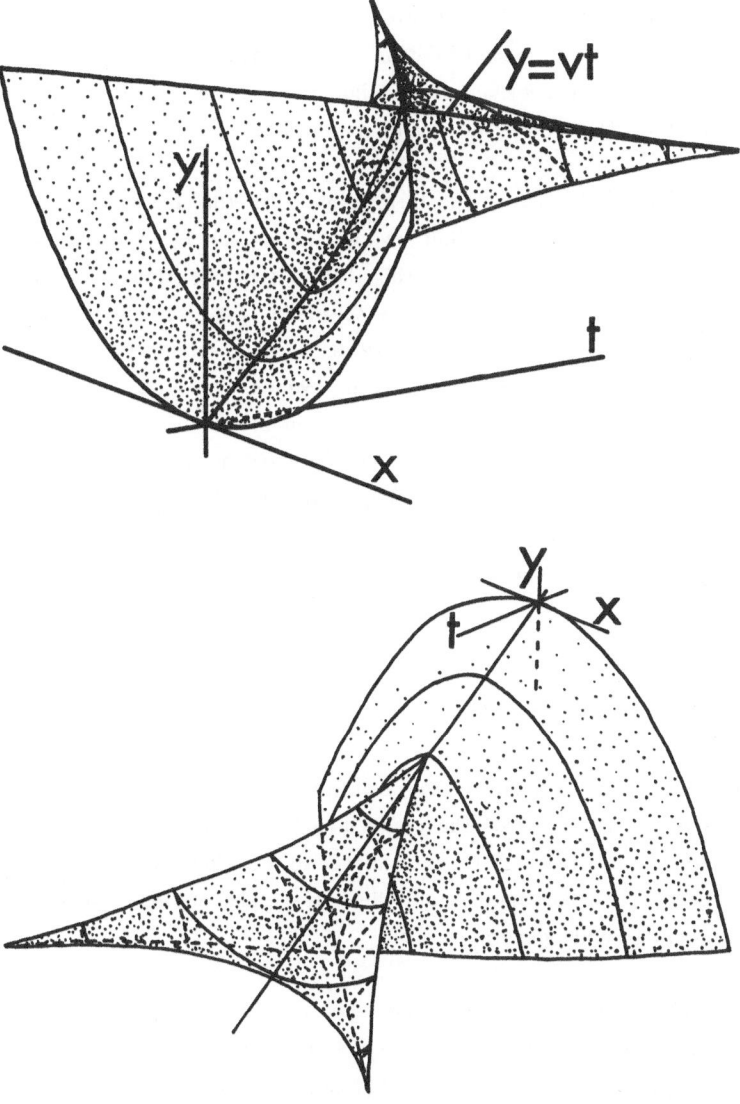

Fig. 4.27: Two views of the modified swallowtail catastrophe. The swallowtail sits on a line y = vt. The sections y = constant through this catastrophe set are the recorded reflections near a concave surface element.

fore, turns out to be a map $R^3 \to R^2$ or $(x,y,t) \to (u,t)$. From the caustic we know that it is a stable pattern in the (x,y)-plane. In addition, we know that the images of the sections through the swallowtail need to have stable positions inside the caustics (Fig. 4.27). A change of the sonic velocity, therefore, cannot alter these spatial patterns, it can only affect the recorded traveltime, i.e. the spreading velocity of the wave front. For the traveltime record this means that the time-axis is stretched or compressed. In the space-time coordinates the sonic velocity can only affect the time-axis. The only allowed transformation of the catastrophe set (Fig. 4.27) by a change of the velocity is, therefore, pure shear in the (y,t)-plane with equation $t = ay$. This transformation does not alter the spatial coordinates of the sections through the modified swallowtail (4.30)), i.e. their projections onto the spatial (x,y)-plane.

How does the local surface pattern map onto the traveltime record? To study this, one has to section the modified swallowtail (4.30) by a plane $w = y = $ constant (in the equations (4.30) 'w' means depth). This gives the image of the local surface in the space-time coordinates, in the $(u,vt/2)$-plane. Fig. 4.27 gives two views of this modified swallowtail, which have been sectioned by a plane $w = $ constant. Again the catastrophe approach summarizes the patterns, which can arise from a local concave reflector area in a very condensed way. By comparison of the observed record with plane sections through the three-dimensional catastrophe set, one can get reasonable qualitative information about the local surface structure. Especially the 'hyperbolic reflections' turn out to represent local surface inversions, which are related to the wave fronts, which have entered the local caustic.

4.2.6   The Traveltime Record as a Plane Map

A second approach to analyze the relation between the local reflector geometry and the traveltime record is versus the plane map $(x,y) \to (u,vt/2)$, which has been defined by equations (4.11). This method is very close to FLOOD's (1980) study of 'hyperbolic reflections' in deep sea echograms. Again the catastrophe approach versus local properties of the reflector will provide general results.

First, one has to specify the mapping equations (4.11). To introduce depth explicitly, the reflector line is locally approximated by a parabola $f(x) = a + bx^2$ like in equations (4.25). The Taylor expansion of 'r' (equation (4.25)) up to order 4 gives the local map

$$u = (1+2ab)x + 2b^2x^3$$
$$r = a + b(1+2ab)x^2 + 2b^3(1-ab)x^4. \tag{4.35}$$

Although this map is very similar to the evolution equation of the wave fronts (4.30), it is not possible to transform it into the standard form of the swallowtail (4.34) by means of simple transformations, which preserve the topological structure. Indeed, as follows from the previous discussion, we should expect arbitrary sections through the swallowtail rather than its standard form.

Now, instead of r we can use $r^2 = v^2t^2/4$ as the distance measurement between the source and the reflection point. The square r^2 is a monotonous function of r because r > 0 (Fig. 4.20). This transformation is not unusual to a seismologist (e.g. KERTZ, 1969), and it allows to formulate the distance r as

$$r^2 = (f(x)^2 + (u-x)^2$$
$$= a^2 + (1-2ab)x^2 + u^2 - 2ux + b^2x^4.$$

(4.36)

This equation can be rewritten as a 'catastrophe potential' if $b \neq 0$,

$$V = (r^2 - a^2)/b^2$$
$$= x^4 + \frac{(1-2ab)}{b^2}x^2 - 2Ux + U^2$$

(4.37)

or

$$V = x^4 + 2vx^2 - 2Ux + U^2$$

with obvious parameter identifications.

The first derivative of this 'catastrophe potential' defines U:

$$V_x = 0 = 4x^3 + 4vx - 2U,$$

(4.38)

i.e. the original cusp catastrophe (eq. 4.20).

This catastrophe potential does not appear in Thom's list of elementary catastrophes, but he discusses it as a selfreproducing singularity or as the stopping potential of the cusp catastrophe (THOM, 1975). In terms of catastrophe theory this potential is the universal unfolding of the cusp catastrophe, and we can embed it into a local potential

$$V_1 = x^5/5 + vx^3/3 + ux^2/2 + u^2x$$

(4.39)

by a proper choice of the parameters. This is a swallowtail with a degenerated parameter space. The parameters 'c' and 'b' from equations (4.33) are now related by $b = c^2$. The critical set appears in the (V,U,v)-space (Fig. 4.28), and the traveltime record is related to sections v = constant through the critical set. Thereby one has to keep in mind that V means r^2, not r. The sections v = constant have locally a swallowtail-like appearance, but, in addition, they have two maxima where the curves bend down again (in the representation of Fig. 4.28).

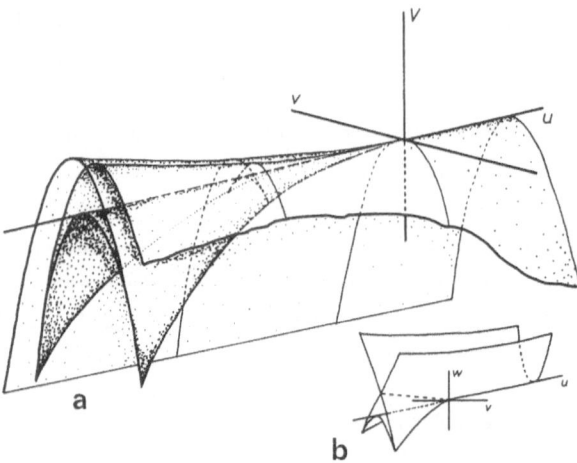

b

Fig. 4.28: The stopping potential of the cusp catastrophe (a) in the parameter space (V,U,v). The positive V-axis is drawn downward for the convenience in comparing it with the standard swallowtail (b) and the hyperbolic reflection of the traveltime record.

The appearance of the two additional maxima above the point of selfintersection in Fig. 4.28 needs an explanation because we cannot expect this pattern from the simple parabolic approximation of the reflector. Similar patterns can be found in the simulated record of Fig. 4.21, but it will turn out that these patterns are of a very different type because they are really related to the surface structure. What happens with the stopping potential, illustrates Fig. 4.29. There, the parabolic reflector line extends over the track line (S). Now, one can construct the image of this abstract surface on the traveltime record in a very simple way. One has just to project the length of the rays, which connect the receiver with the reflection points straight downward from the point where they intersect the track line. This gives the curve (r), i.e. the image on the traveltime record. This construction can also be done for those parts of the reflector line which extend above the track line. Because traveltime is measured without a directional component, i.e. it can only assume positive values, the curve (r) bends down again, as one moves away from the intersection point of (S) and (r). Therefore, one has to choose

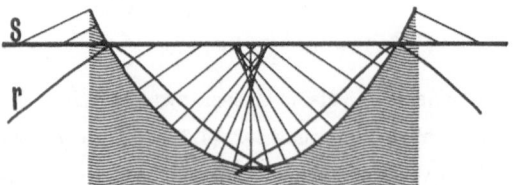

Fig. 4.29: The abstract situation that the track line (S) intersects the reflector line. In this case, the traveltime record (r) reaches the track line at the intersection point and bends then down again because r can assume only positive values.

carefully the correct interval if the stopping potential is used as a model for the travel-time record. The correct interval is, in any case, located between the two maxima of the sections v = constant of Fig. 4.28.

If one analyzes the critical surface of Fig. 4.28 with the noted restrictions in mind, then it turns out that the typical 'hyperbolic reflections' with a swallowtail-like appearance are restricted to a limited range of the parameter v. If v is positive, one has a convex reflector, which in a topological sense is recorded correctly. As v assumes sufficiently large negative values, the 'hyperbolic reflections' turn smoothly into a more parabolic appearance, which, like the 'hyperbolic reflections', is an inversion of the local reflector topology -- a concave surface element turns into a convex image. Those 'parabolic reflections' are also well known from echograms (FLOOD, 1980), but, more common-ly, they are found within basement reflections (Figs. 4.14, 4.15).

As was shown in the last section, the approach versus wave fronts provides another frame to summarize the images on the traveltime record. The advantage of the plane map approach is that the 'stopping potential' represents the images in a still more con-densed way.

4.2.7 Singularities on the Reflector Line

So far, a very simple reflector model was used. In the case of faults, folds and flexures the situation may become more complicated although a local parabolic approxi-mation with rotation of the coordinate axes may be still possible. The most simple case, where one can find such a critical situation, are flexures and folds. A first impression

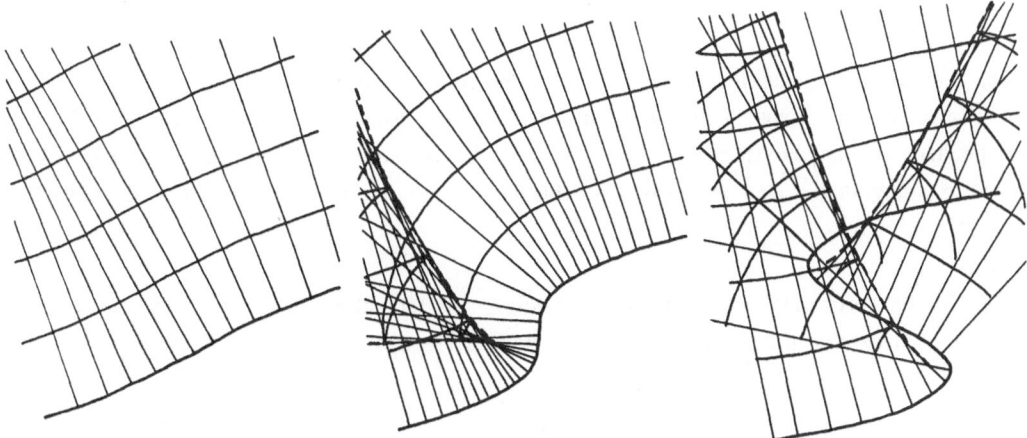

Fig. 4.30: First order approximation of ray systems and wave fronts near flexures.

of what may happen near a fault gives the simple linear model of section 4.2.2 (Fig. 4.23). What is actually new in this linear approximation, is the appearance of a shadow zone. A flexure can be simulated by a cubic equation $x = y^3 + ay$ which also includes simple folds. Fig. 4.30 gives a rough approximation of rays and wave fronts which arise from the cubic reflector line model with $a > 0$, $a = 0$ and $a < 0$. For $a < 0$, the caustic patterns can be approximated by a parabolic approximation at the extrema of the cubic equation, but only parts of the wave fronts scatter back to the track line, i.e. only one branch of the caustic intersects with the track line. Fig. 4.31 tries to capture the behavior of the caustic over a family of cubic reflector lines. For compari-

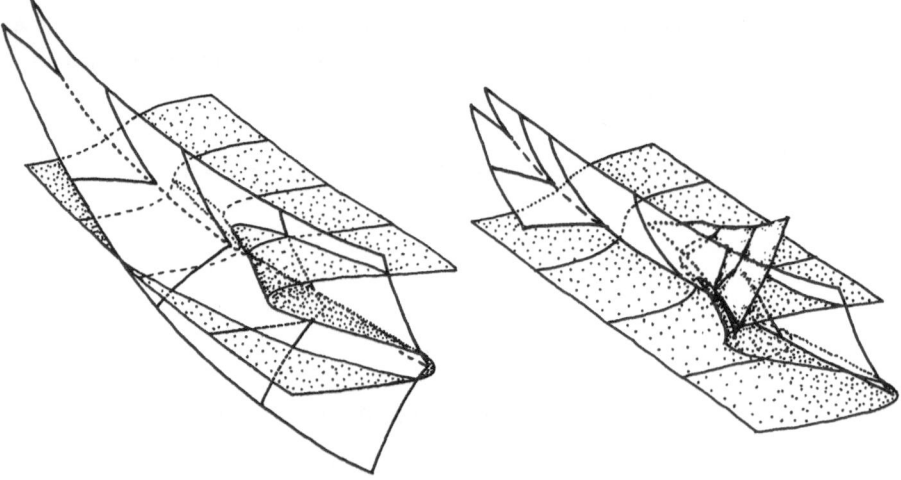

Fig. 4.31: The caustics of a family of cubic reflector lines. Left: The family of caustics of only one extremum (the lower one). Right: Separation of the caustics into their relevant parts, i.e. the branches which reach the track line.

son, the family of caustics for only one extremum is also shown. These graphical methods only give a very rough idea of what happens near such structures, but it is not the scope here to analyze these problems in detail. In this context it becomes at least necessary to study diffraction patterns. For this approach see DANGELMAYR & GÜTTINGER (1982).

Similar problems arise if the reflector has singular points like the cycloid of Fig. 4.18 in section 4.2.1. The cycloid can be described by the map

$$x = t - \sin(t)$$
$$y = 1 - \cos(t). \tag{4.40}$$

By taking a Taylor expansion near the cusp point, one finds

$$x = t^3$$
$$y = t^2$$

(4.41)

where the constants have been absorbed in x and y for convenience. If the parameter t is eliminated, one finds that the cusp point equals the semicubic parabola $y^3 = x^2$. The main point, however, is not that we have a cusp, but that the singularity is an isolated point of the reflector line at which dx/dt = 0 and dy/dt = 0. The caustic near the critical point is given as the loci of the radii of curvature on the normals of the semicubic parabola:

$$x_c = 4t^3 + \frac{4}{3}t$$
$$y_c = \frac{9}{2}t^4 - t^2.$$

(4.42)

Thus, not even too close to the isolated singular point (t = 0) the caustic behaves like the map

$$u = t^2$$
$$v = t,$$

(4.43)

i.e. it is a fold catastrophe (Fig. 4.32; LU, 1976). The term 'not too close' means that t^3 is much smaller than t and that t^4 is much smaller than t^2. At a fold caustic the rays are only located on one side of the caustics and cause, therefore, locally a shadow zone. A detailed analysis of such singular points on the reflector line would require a topological classification, and it would be necessary to study the wavefield rather than the ray system.

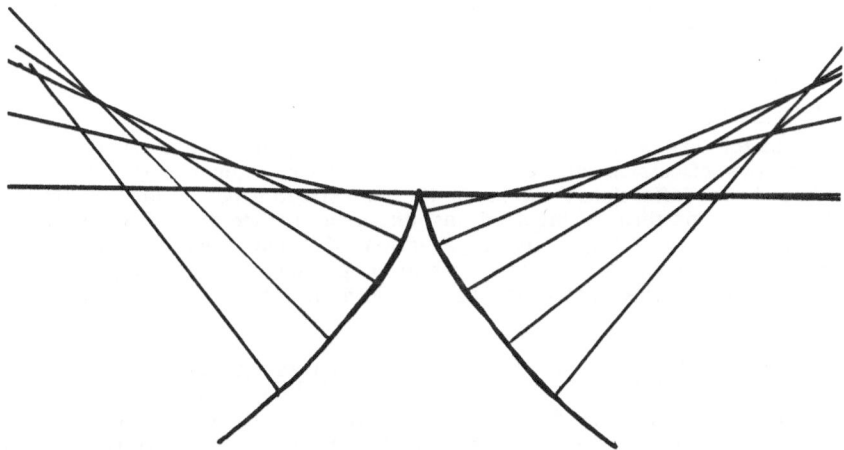

Fig. 4.32: The fold catastrophe (caustic) near a singular point on the reflector causes a shadow zone.

Table 4-1: Summary of the Ray Model

The traveltime record in its most critical case corresponds to sections y = constant (y:depth) through a swallowtail catastrophe which is located on a line y = vt in the three-dimensional space (x,y,t). The various types of specialized deformations depend on the local curvature of the reflector line, on the distance between the track line and the critical point on the reflector line, and on the sonic velocity of the medium. In the parameter space depth of the critical point (a) and local curvature (b), the critical boundary line for an image inversion, i.e. for the occurrence of 'hyperbolic reflections', is given by the hyperbola 1 - 2ab = 0. This hyperbolic equation simply compares the local curvature of the reflector with a circular wave front at depth 'a'. In detail, one finds that these parameters affect the traveltime record in the following way:

I) Spatial patterns, the cusp catastrophe

1) The local curvature of the reflector line:

Only convex areas of the reflector line are spectacular (cause trouble within the record) because a cuspoid caustic evolves. Two special situations occur:

a) The local approximation of the reflector line requires a rotation of the local coordinate system with respect to the global one. The sections through the catastrophe set becomes oblique. This pattern can be detected on the traveltime record because the 'hyperbolic reflections' are asymmetric.

b) Different local curvatures (b = k/2) or the reflector cause a dislocation and stretching (compression) of the caustic in the spatial coordinates. This deformation can only be distinguished from (2) if the true depth position of the spectacular point on the reflector line is known.

2) The height of the track line above the reflector line:

Because the reflection pattern depends on the relation between the curvature of the incident wave front and the curvature at the spectacular point on the reflector line, this case cannot be distinguished from a change of the local curvature of the reflector without additional information (e.g. a measurement of true depth). This parameter chooses a special line through the catastrophe set of the cusp which is stably located in the space coordinates. Because the cusp catastrophe is the bifurcation set for the swallowtail and the discussed stopping potential, this parameter also appears in the other catastrophes.

3) Extrema of curvature:

In case the reflector has a local minimum of curvature, it can be approximated by a parabola, and the discussion of sections 4.2.1-7 holds: Typical patterns inside the caustic are 'hyperbolic reflections'. However, if the reflector has a local maximum of curvature, the caustic pattern is inversed, as discussed in section 4.2.8. Anyway, the previous discussion remains valid if the propagation of wave fronts is inversed. After the wave fronts have passed through the caustic, a parabolic reflection pattern results which allows to distinguish this case from the 'standard situation'.

II) Space-time patterns: the swallowtail catastrophe

The sonic velocity of the medium does only affect the traveltime. This parameter can, therefore, cause only those transformations which let the space pattern invariant -- pure shear in the (y,t) -- plane. The catastrophe set, which describes the evolution of the wave fronts, is a modified swallowtail which is located on a line y = vt. The traveltime images are plane sections through this catastrophe set. Alternatively, the traveltime image can be described by the unfolding of the cusp catastrophe, i.e. by its stopping potential. The latter approach gives a description in the coordinates (x,y,v^2t^2).

Table 4-2: Summary of strategies in the analysis of traveltime records	
"wave front approach"	"plane mapping method"
Construction of the <u>ray system</u> (normals of the local reflector element)	

The <u>caustic</u> or the envelope of the rays (centers of curvature)	The catastrophe map along the track line, the <u>cusp catastrophe</u>

Evolution of the wave fronts along the rays, the <u>swallowtail catastrophe</u>

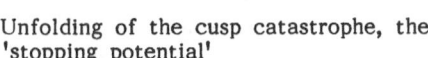

traveltime sections through the catastrophe set of the wave fronts -- the <u>modified swallowtail</u>	Unfolding of the cusp catastrophe, the <u>'stopping potential'</u>

The local image of the traveltime record

4.2.8 Generalized Reflector Patterns in Two and Three Dimensions

In case the reflector can be described by an explicit function $y=f(x)$, the previous discussion provides a finite classification of reflector patterns as long as a linear ray model is sufficient and catastrophe theory provides a frame for this classification, as summarized in tables 4-1 and 4-2. However, the application of catastrophe theory requires local coordinate changes, which sometimes may be assumed inadequate for the problem. In the previous discussion it turned out that the traveltime record depends on a parameter $s=1-2ab$ which appears in all equations -- for the caustic, the wave fronts and the traveltime record. The parameter 'a' is equivalent to the depth of the reflector, and '2b=k' is its local curvature (cf. equation 4.13). The parameter 's', therefore, provides a simple interpretation, it measures the relation between an incident wave front with radius 'a' (depth) and the radius of curvature of the reflector. Image inversion occurs for $a > 1/(2b)$, i.e. if the radius of the incident 'wave front' is larger than the radius of curvature, multiple reflections arise locally because the curvature of the reflector increases, as one departs from the critical minimum. Fig. 4.33 illustrates this viewpoint.

A) The Deformed Circle and the Dual Cusps

A natural question is what happens if the reflector has a different structure, i.e.

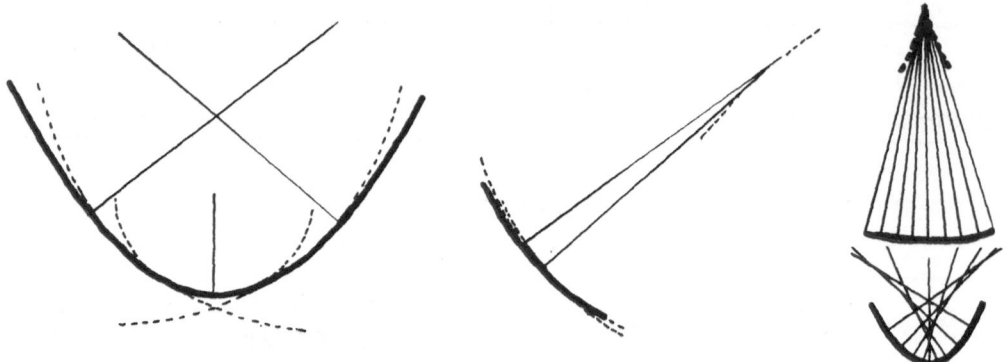

Fig. 4.33: The contact between the incident wave front and the circle of curvature determines the possible number of received reflections: In the case of a parabolic reflector, multiple reflections result only if the curvature of the incident wave front is larger than the local curvature of the reflector, i.e. if the shotpoint is located inside the 'caustic' of normals. The usual situation is a fold point on the caustic (b); a cusp point appears only at a local extremum of curvature.

if the curvature decreases, as one departs from the minimum. This causes a different type of contact between the circle of curvature and the reflector: The reflector is totally bound to the convex side of the circle of curvature, a situation which cannot arise in the case of a locally 'parabolic reflector'. An appropriate way to study both situations simultaneously is to consider a perfect circular arc, and to transform it by a simple affine transformation

$$\begin{pmatrix} x \\ y \end{pmatrix} = r \begin{pmatrix} 1 & 0 \\ 0 & e \end{pmatrix} \begin{pmatrix} \cos\psi \\ \sin\psi \end{pmatrix} \tag{4.44}$$

which takes the circle into an ellipse. Fig. 4.34 illustrates how the reflector element, its contact with the circle of curvature and the caustic are altered by a smooth change of the parameter 'e':

In the case $0 < e < 1$ the ellipse has a local minimum of curvature. The circle of curvature is bounded to the concave side of the reflector, which, therefore, can be approximated by a parabola, and the previous discussion can be applied.

For e=0, the reflector is a perfect circular arc. All rays pass through a single point -- a singularity with indefinite codimensions. This situation is structurally unstable, as any small disturbance transforms the singular point into a caustic.

If $e > 1$, the circle of curvature is located on the convex side of the reflector, and a new pattern arises. However, the caustic is again cuspoid, similar to the caustic of a cycloid (Fig. 4.18); but the cusp points into the opposite direction than in the case of a parabolic reflector.

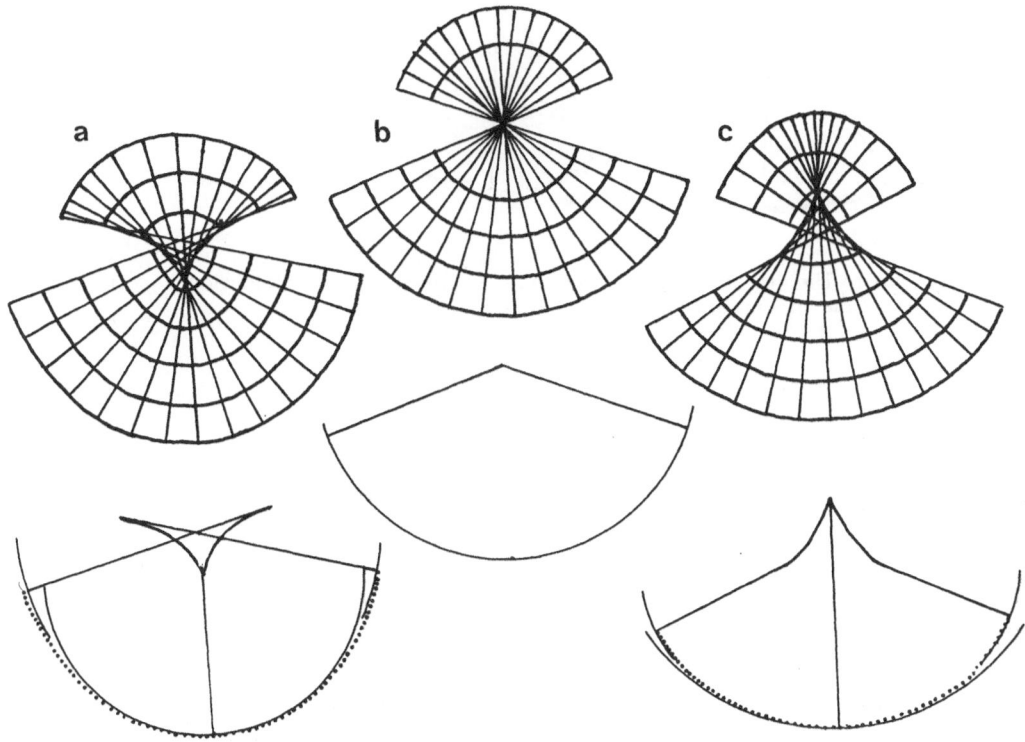

Fig. 4.34: Rays and wave fronts from an elliptic reflector. a: $0 < e < 1$, b: e=0, c: $e > 1$. See text for discussion.

The type of caustic thus depends on the type of contact between the circle of curvature and the reflector. The ellipse still provides a rather special example. A more general viewpoint and classification can be derived if the arguments of section 4.2.2 are applied to more general curves.

Locally, the circle of curvature provides a rather good approximation of a two--dimensional curve. Choosing its center as the origin of a polar coordinate system we can describe the reflector by an equation

$$r = R + f(\theta),$$
(4.45)

where R is the local radius of curvature and $f(\theta)$ describes the deviation of the curve from the perfect circular arc (cf. Fig. 4.35). The question is what we can infer about the function $f(\theta)$. The radius of curvature in polar coordinates is given by

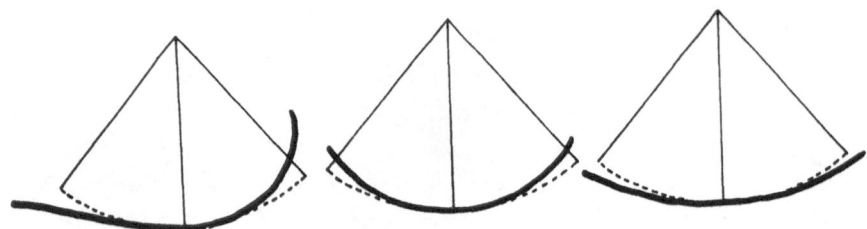

<u>Fig. 4.35:</u> The three possible contacts between a two-dimensional reflector and its circle of curvature.

$$R = \frac{(r^2 + r'^2)^{3/2}}{r^2 + 2r'^2 - rr''} .$$

(4.46)

At $\theta = 0$ the reflector has curvature R, and this is the case if f(θ) satisfies the three conditions f(0)=0, f'(0)=0, and f''(0)=0 as can easily be verified from the standard equation (4.46). If we use a power series to approximate f(θ), then this series cannot involve powers less than three, i.e. we need at least a function $f(\theta) = \theta^3 +...+$higher terms. Such functions, of course, are really flat at the origin, their curvature vanishes at $\theta = 0$.

However, $f(\theta) = \theta^3$ is an odd function, and if we insert it into equation (4.45), it becomes clear that the circle of curvature intersects the reflector in some neighborhood of $\theta = 0$; the local reflector model is a 'spiral arc' with monotonously increasing (decreasing) curvature in a sufficiently small neighborhood of $\theta = 0$ (Fig. 4.35a). A sign change of the leading term ($f(\theta) = \pm\theta^3$) simply reflects the intersection pattern at the ray $\theta = 0$; the pattern, however, does not change.

The situation becomes different if the power series starts with a fourth order term. Then the reflector deviates symmetrically from the circle of curvature, and a sign change of the leading fourth order term changes the type of contact: For $+\theta^4$ the circle of curvature is entirely on the concave side of the reflector while for $-\theta^4$ it is on the convex side (Fig. 4.35).

The two alternative power series with leading terms of order three or four are really distinct and exclude one another, as now will be shown. A local reflector approximation involving both terms could always be brought to the form

$$f(\theta) = \theta^3 + a\theta^4 + ... + \text{higher terms}.$$

(4.47)

However, by a redefinition of the zero angle $(\theta - \theta - \frac{1}{4a})$,equation (4.47) can be transformed into

$$\theta^4/(4a) - \frac{3}{2}\theta^2 + 2a^2\theta + (a^4 - a^3).$$ (4.48)

In equation (4.48) the radius of curvature is given by $(R+c)$, and $f(\theta)$ has again to satisfy $f(0)=f'(0)=f''(0)=0$, i.e.

$$\frac{1}{a}\theta^3 - 3\theta^2 + 2a^2 = 0$$

$$\frac{3}{a}\theta^2 - 6\theta = 0 .$$ (4.49)

These two equations, however, are usually not zero, and the function $f(\theta)$ is dominated by the first and second order terms with non-vanishing first and second order derivatives and, therefore, does not satisfy the requested approximation.

Therefore, our problem is, locally, strongly equivalent to a power series which starts either with a third or a fourth order term, and catastrophe theory implies a fold or cusp catastrophe. The critical point in our problem is the point r=R, the center of the circle of curvature which, of course, is a point on the evolute of the rays, i.e. a point on the caustic. Sufficiently close to $\theta =0$, the radii of our polar coordinate system coincide with the rays. The transformation $\rho =r-R$ maps the reflector (the wave front) to the critical point. Near this point, we take the reflector as $f(\theta)=\theta^4$ or more conveniently, we use the unfolding

$$\rho = \pm\theta^4/4 + u\theta^2/2 + v\theta.$$ (4.50)

We cannot choose u and v freely because $f(\theta)$ has to satisfy $f'(0)=f''(0)=0$. This leads to the set of equations

$$v = \mp\theta^3 - u\theta$$

and (4.51)

$$u = \mp3\theta^2.$$

If we solve for u and v in terms of θ and insert this in equation (4.50), this equation simplifies to a fourth order term as required. However, if we use u and v as local orthogonal coordinates, then we can eliminate θ and find one of the dual cusps

$$(\frac{u}{3})^3 = \mp(\frac{v}{4})^2$$ (4.52)

i.e. the caustic we expect. In a spatial interpretation, u is the (negative) first, v the second derivative of the function ρ . Interpreted as vectors, they provide a local orthogonal frame and capture qualitatively the dislocation of rays close to the critical point r=R. Similar arguments can be applied to the case $f(\theta) = \theta^3$, the critical points are fold points.

If we restrict our attention to local structures, there is not more than fold and cusp points on a caustic. Their occurrence is a function of the contact between the reflector and its local circle of curvature, as illustrated in Fig. 4.35. In terms of reflection patterns, however, 'local' is rather relative. In this context, a fold point is a point where two rays intersect; however, this is only the case on the caustic itself. In the interior of a caustic (cf. Fig. 4.34), which is not related to a singular point on the reflector (e.g. Fig. 4.32), we find that three rays intersect at every point. Thus, fold points are not such important from a less local viewpoint. Important, however, is the difference between the dual cusps because they provide an essential source for the seismic interpretation.

In one sense the dual cusps are not different, they are simply dual reflections at the x-axis which result from a sign change of the leading power terms, and thus are all related patterns. This is obvious because any wave front can be considered as a reflector and vice versa -- here a wave front is a map of the reflector along the rays preserving angles. The wave front patterns of the two dual cusps, therefore, are identical, only the direction of propagation is inversed. This is a nice result because it shows that the previous discussion holds also for the dual cusp if the direction of wave propagation is inversed; and the earlier discussion provides really a catalogue of the essential reflection patterns as far as a linear approach is sufficient.

On the other hand, there remains a difference between the dual cusps. In the case of a locally parabolic reflector, the wave fronts are sections through the swallowtail with its cusps and selfintersections, and the traveltime records in the critical case are

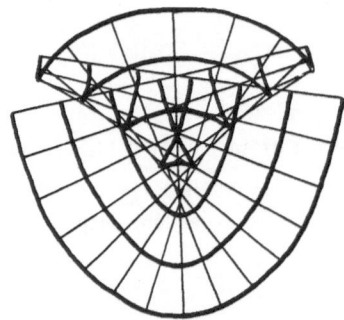
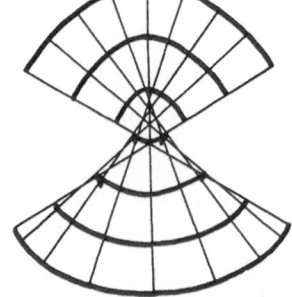

Fig. 4.36: Wave fronts of the dual cusps.

'hyperbolic reflections', again with cusps and selfintersections. In the case of the dual cusp, the reflector is an elliptic arc which is bounded to the interior of the cuspoid caustic; as soon as the image passes through the cusp point, the wave fronts have a 'parabolic' appearance, and thus has the traveltime record; cusp points and selfintersections then are missing. A typical pattern, which commonly arises, is a series of parabolae which alternatively correspond to synclines and anticlines, and which intersect on the traveltime record, but without cusp points. In the case the track line sections the caustic, swallowtail patterns may arise, but they are inverted with respect to the patterns arising from a 'parabolic reflector' (Fig. 4.36).

In summary, the various reflection patterns, which may arise, are well classifiable in terms of the contact between the reflector, its circle of curvature and the incident wave front (distance from the source). Usually there should be enough information available for a qualitatively correct interpretaion. The linear ray model, of course, is only a first approximation, but the principal relationships remain stable even if the sonic velocity of the medium is not a constant.

B) Three-Dimensional Patterns -- The Double Cusp

At least, a few remarks shall be made in what respect the simplified model of linear rays and especially of a two-dimensional reflector line gives insight into a larger class of images which may result from complicated reflector topologies. The two-dimensional approach extends without difficulties to cylindrical surface elements or, more generally, to parabolic surface points. Fig. 4.37 gives two examples -- a cylindrical and a conical surface -- that show how the caustic and a single wave front are located over the surface. In such cases, the traveltime record will depend on the relation between the axis of the syncline and the track line -- one may find 'hyperbolic reflections', onlapping patterns, doubled or tripled reflections (Fig. 4.38). Thus, an irregular topography, which impresses its structure onto the wavefield, can cause nice multiple reflection patterns which look like perfectly stratified sediments; and, therefore, one may ask how much onlapping features in Fig. 4.15 are real, and which ones are due to the rough topography of the basement. The complexity of these effects increases if one considers farfield effects or more complicated surface elements like hyperbolic and elliptic surface points. Representing the surface near (x_o,y_o,z_o) by $z=f(x,y)$ the evolution equation for the wave fronts becomes

$$\{(u,v,w) \in (x,y,z) \mid ((x-u)^2 + (y-v)^2 + (w-f(x,y))^2)^{1/2} = r = \text{const.} \}. \tag{4.53}$$

In the case of a parabolic or hyperbolic surface point, the family of rays is given by the (vector) equation

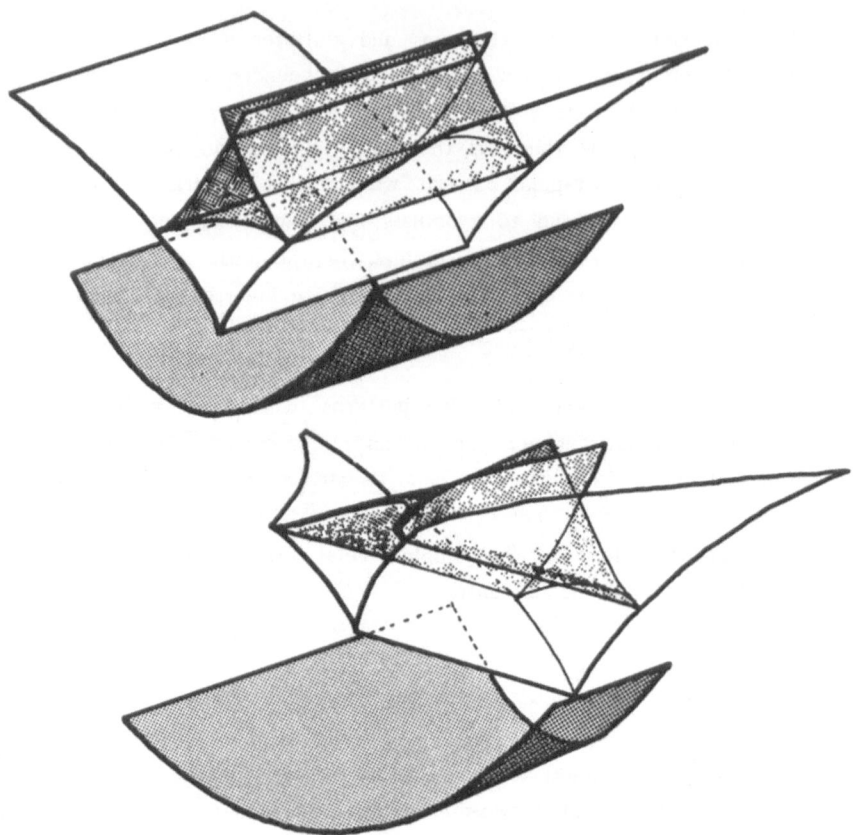

Fig. 4.37: The caustic and a single wave front over a cylindrical (above) and conical (below) surface.

$$r = (x,\ y,\ x^2 \pm ay^2) + \lambda(2x,\ \pm 2ay,\ -1), \tag{4.54}$$

and a point on the track line may be given as (x_o, y_o, z_o). To see, which surface points map onto the track line, one has to solve the equation

$$(x_0,\ y_0,\ z_0) = (x,\ y,\ x^2 \pm ay^2) + \lambda(2x,\ \pm 2ay,\ -1). \tag{4.55}$$

Let the track line be located at z_o, then by elimination of the parameter λ , one finds the relationship

$$\lambda = x^2 \pm ay^2 - z_0$$

$$x_0 = (1-2z_0)x + 2x^3 \pm 2ay^2 x$$

$$y_0 = (1 \mp 2z_0 + 2ay^3 \pm 2x^2 y, \tag{4.56}$$

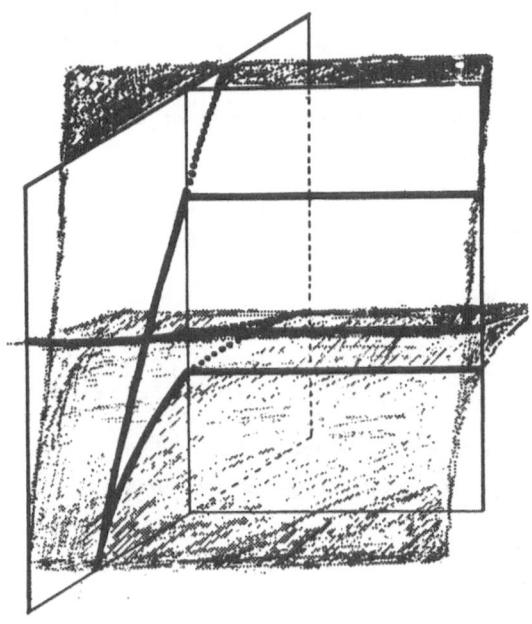

Fig. 4.38: Sketch of the traveltime record of a cylindric syncline with track line sections parallel and oblique to the syncline axis.

a map which is a special degenerated case of the double cusp catastrophe including the standard umbilic catastrophes. The caustic patterns, which result from the double cusp, are rather complicated. A full discussion of three-dimensional phenomena is above the scope of this discussion; however, a detailed study in terms of standard catastrophes was given by DANGELMAYER & GÜTTINGER (1983).

4.2.9 Distributed Receivers

Seismic shooting rarely resembles the idealized situation that source and receiver are at the same place. However, as turned out during the previous discussion, the results found from rather idealized assumptions hold for a much wider class of 'disturbed' problems. It will be shown here that the principal results still hold if source and receiver are at different places, or if a chain of receivers is used. In the latter case, not a single reflection but the reflected and deformed wavelet is recorded. What we shall do here is, therefore, to study how the reflected wavelet deforms.

The reflection of a wavelet is governed by Snell's law of equal angles, i.e. the angle an incident ray forms with the normal of the reflecting surface is the same as the angle the reflected ray forms with the same normal. A convenient way, therefore, is to view the incident and the reflected rays in terms of the reflector. Let the reflector be given in terms of its local curvature, i.e. with the center of the global coordinate system at the center of its local circle of curvature (cf. equ. 4.45): Locally the reflector can be written

$$\begin{bmatrix} x \\ y \end{bmatrix} = r \begin{bmatrix} \cos\theta \\ \sin\theta \end{bmatrix}, \tag{4.57}$$

and the normal rays are

$$\begin{bmatrix} x_n \\ y_n \end{bmatrix} = r \begin{bmatrix} \cos\theta \\ \sin\theta \end{bmatrix} + \lambda \left(-r \begin{bmatrix} \cos\theta \\ \sin\theta \end{bmatrix} + r' \begin{bmatrix} -\sin\theta \\ \cos\theta \end{bmatrix} \right). \tag{4.58}$$

Now, in a plane problem we can express the incident rays in local coordinates by means of the tangent (t) and normal (n) vectors at the surface:

$$r_i = r + \lambda(-\alpha n + \beta t), \tag{4.59}$$

and the reflected rays are simply the reflections of incident rays at the normals

$$r_r = r + \lambda(-\alpha n - \beta t). \tag{4.60}$$

The coefficients 'a' and 'b' can be determined to satisfy special conditions of the source, e.g. in the case of a point source, equation (4.59) leads to a pair of linear equations from which the coefficients can be determined. A very simple system arises if the reflector is locally a perfect circular arc. The equations for the incident and reflected rays then simplify to the pair of equations

$$\begin{bmatrix} x_i \\ y_i \end{bmatrix} = (r-a\lambda) \begin{bmatrix} \cos\theta \\ \sin\theta \end{bmatrix} + \lambda b \begin{bmatrix} -\sin\theta \\ \cos\theta \end{bmatrix}$$

$$\begin{bmatrix} x_r \\ y_r \end{bmatrix} = (r-a\lambda) \begin{bmatrix} \cos\theta \\ \sin\theta \end{bmatrix} - \lambda b \begin{bmatrix} -\sin\theta \\ \cos\theta \end{bmatrix}. \tag{4.61}$$

The condition that the incident rays originate from a point source requires that these rays pass through the source point for some value of λ. Without loss of generality, we can choose the value $\lambda=1$, and the values for the parameters 'a' and 'b' can be determined from the linear equations

$$(r-a)\cos\theta - b\sin\theta = x_0$$
$$(r-a)\sin\theta + b\cos\theta = y_0$$

to be

$$a = r - (y_0\sin\theta + x_0\cos\theta)$$
$$b = \qquad y_0\cos\theta - x_0\sin\theta \; . \qquad\qquad\qquad (4.62)$$

Because of the symmetry of the circular arc a simple rotation allows to locate the source formally at $(x_o,0)$ so that the previous equations simplify further. If one inserts 'a' and 'b' from equation (4.62) into equations (4.61), one finds a simplified equation for the incident rays

$$\begin{pmatrix} x_i \\ y_i \end{pmatrix} = r \left[(1-\lambda) \begin{pmatrix} \cos\theta \\ \sin\theta \end{pmatrix} + \lambda \begin{pmatrix} x_0 \\ 0 \end{pmatrix} \right], \qquad\qquad (4.63)$$

and the reflected rays are

$$\begin{pmatrix} x_r \\ y_r \end{pmatrix} = r \left[(1-\lambda) \begin{pmatrix} \cos\theta \\ \sin\theta \end{pmatrix} + \lambda x_0 \begin{pmatrix} \cos2\theta \\ \sin2\theta \end{pmatrix} \right]. \qquad\qquad (4.64)$$

As previously, the caustic of the reflected ray system is of special interest. If we consider equation (4.61) as a map, the caustic is equivalent to its singular set, which can be determined from the condition that the Jacobian of the map vanishes, i.e. that

$$\mathbf{J} = \begin{vmatrix} x_\theta & x_\lambda \\ y_\theta & y_\lambda \end{vmatrix} = x_\theta y_\lambda - x_\lambda y_\theta = 0 .$$

From this condition and equations (4.61) and (4.64) we determine the critical set in terms of λ :

$$\lambda = \frac{a}{2(a^2+b^2)-a} = \frac{1 - x_o\cos\theta}{1+2x_o^2-3x_o\cos\theta} \qquad \text{if } y_o=0 . \qquad (4.65)$$

If we insert these values for λ into equation (4.64), we find an equation for the caustic

$$\begin{pmatrix} x_r \\ y_r \end{pmatrix} = r \left[2\frac{x_o^2 - x_o\cos\theta}{1+2x_o^2-3x_o\cos\theta} \begin{pmatrix} \cos\theta \\ \sin\theta \end{pmatrix} \frac{x_o - x_o^2\cos\theta}{1+2x_o^2-3x_o\cos\theta} \begin{pmatrix} \cos2\theta \\ \sin2\theta \end{pmatrix} \right] \qquad (4.66)$$

which looks rather complicated. However, a simple observation is important. Let us compute the values $(1-\lambda)$ and λx_o at $\theta=0$:

$$(1-\lambda) = 2\frac{x_o^2 - x_o}{1+2x_o^2} \; ; \qquad \lambda x_o = -\frac{x_o - x_o^2}{1+2x_o^2} \; .$$

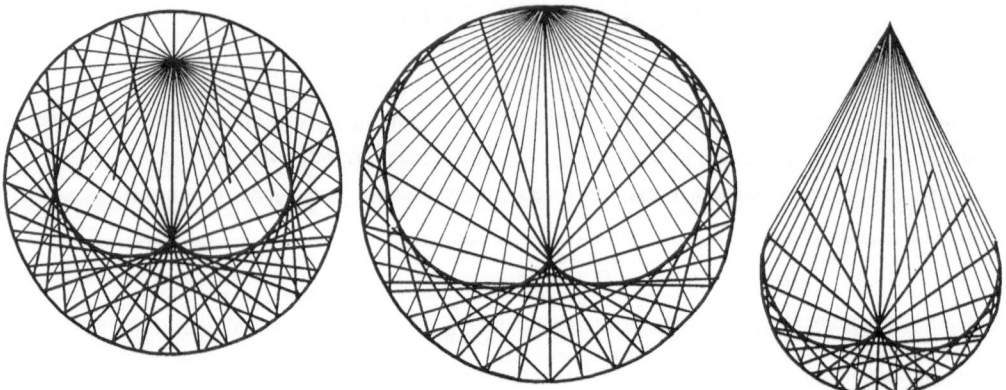

<u>Fig. 4.39:</u> The cardioid caustics of a circular reflector and their relation to point sources.

Locally, near $\theta = 0$, we have the simple relationship $2(1- \lambda)= \lambda x_o$, and this means that near this special point the caustic behaves like a cardioid independent of the complexity of our original equation. The cardioid, however, has a cusp point at $\theta = 0$, and this is a standard cusp point, as can easily be shown by developing the equations

$$x = r(2\cos\theta - \cos2\theta); \qquad y = r(2\sin\theta - \sin2\theta)$$

in Taylor series near the critical point $\theta = 0$:

$$x \sim 1 + \theta^2; \qquad y \sim \frac{10}{9}\theta^3 \quad ---> \quad (x-1)^3 = (\frac{9}{10} y)^2.$$

What we now can do with the source point, is to dislocate it along the x-coordinate (Fig. 4.39). Clearly, a critical situation arises if the source is located at (0,0), the center of the circle of curvature. In this case, all rays pass through the origin, the caustic degenerates to a singular point, and we would not receive any reflections at points besides this degenerated singularity.

If $0 < |x_o| > R$, we find that the caustic has formally two cusp points if we consider not simply a circular arc but a full circle. These cusp points are given by $\theta = 0$ and $\theta = \pi$. In addition, we observe that these cusp points are simple inversions of the corresponding source locations $x_o \rightarrow -x_o$. A somewhat striking point is that we always have the same type of a cusp (what we called the dual cusp of the reflection problem) independent of the radius of the incident wavelet. The caustic pattern, therefore, does not depend on the contact between the reflector and the (circular) incident wavelet.

Another special situation occurs if $|x_o| = R$. In this case, we have only one cusp point, the other one degenerates into a fold point with its tangent coincident with the tangent of the circular reflector -- the caustic becomes a perfect cardioid. In the case the source point is located outside the circle, there remains only one cusp point, but in addition we find two critical fold points where the deformed cardioid has tangential contact with the circular reflector. In some sense, the situation $x_0=1$ defines a bifurcation point. However, if $x_0 \to \infty$, we find again a symmetric solution, the caustic is now a nephroid (cf. POSTON & STEWART, 1978) whereby the symmetry refers to the two sources $x_0=\pm\infty$.

The caustic phenomena associated with a point source and a perfect circular reflector, therefore, can be summarized as continuous deformations of a cardioid. The stable pattern is the cusp point of the cardioid, which locally remains the identical cusp caustic independent of the location of the source. Now, the circular reflector is unstable, and the question arises what happens if it is deformed. Before going in details, we

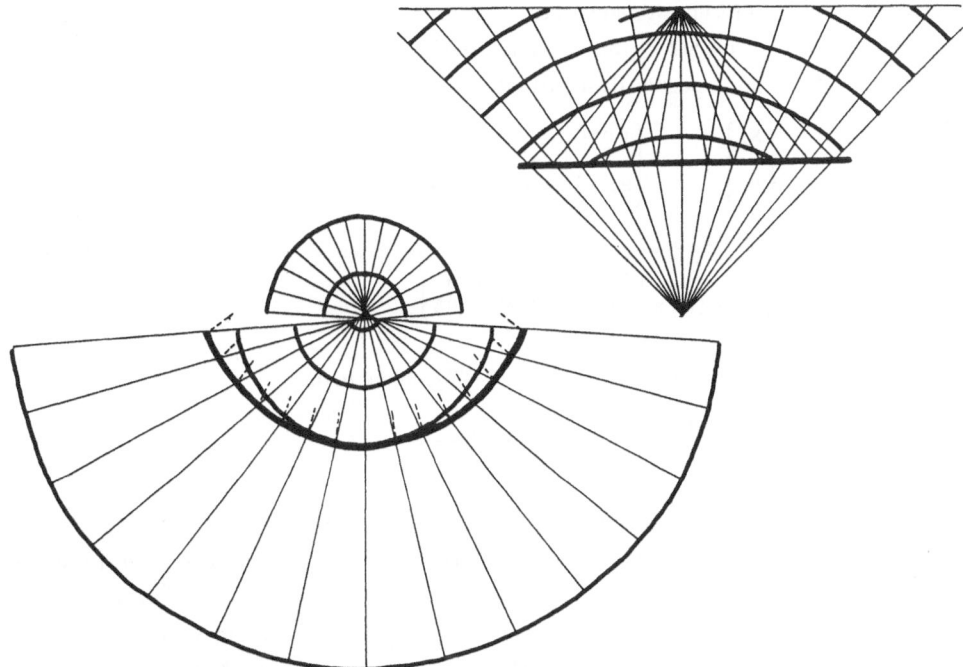

Fig. 4.40: The virtual sources of a planar and circular reflector. The wave fronts provide virtual reflectors.

first observe that there exists a virtual surface, for which the reflected rays of the wavelet are normals. In the case of a plane reflector, this virtual surface is again a point source, a standard example in seismology (Fig. 4.40). In a more general sense, every wave front is a potentially virtual reflector surface because the wave fronts intersect the rays orthogonally. In the case of linear rays, the wave fronts are found from the normalized equation (4.61), i.e. from

$$\begin{bmatrix} x_r \\ y_r \end{bmatrix} = r \begin{bmatrix} \cos\theta \\ \sin\theta \end{bmatrix} + \frac{\lambda}{(a^2+b^2)^{1/2}} \left(-a \begin{bmatrix} \cos\theta \\ \sin\theta \end{bmatrix} - b \begin{bmatrix} -\sin\theta \\ \cos\theta \end{bmatrix} \right). \tag{4.67}$$

The traveltime is equivalent to the sum of the length of the incident and reflected rays. If the receivers are on the same x-level as the source, the traveltime is given by

$$2t = (a^2+b^2)^{1/2}(1 - \frac{x_0 - r\cos\theta}{a\cos + b\sin\theta}), \tag{4.68}$$

and the identical traveltime record would be received in a system where source and receiver coincide, either from a virtual source or a virtual reflector which, of course, is simply a wave front (Fig. 4.40). Now, we can use the discussion of the last section. A critical situation arises if the virtual reflecting surface becomes a circle. Any small deformation then deforms it, and the singular caustic point evolves in either one of the dual cusps. We consider this degenerated situation and disturb the reflected rays (the normals) by a not necessarily constant rotation

$$\begin{bmatrix} x_r \\ y_r \end{bmatrix} = r \begin{bmatrix} \cos\theta \\ \sin\theta \end{bmatrix} - \lambda r A \begin{bmatrix} \cos\theta \\ \sin\theta \end{bmatrix} + \lambda r' A \begin{bmatrix} -\sin\theta \\ \cos\theta \end{bmatrix} \tag{4.69}$$

where

$$A = \begin{bmatrix} \cos\alpha(\theta) & -\sin\alpha(\theta) \\ \sin\alpha(\theta) & \cos\alpha(\theta) \end{bmatrix}. \tag{4.70}$$

The critical set can again be found from the Jacobian to be

$$-\lambda = \frac{(\rho^2+\rho'^2)\cos\alpha}{(1+\alpha')\rho^2 \ (2+\alpha')\rho'^2-\rho\rho''}. \tag{4.70}$$

The deformation of the original ray system, therefore, consists of a rotation as defined by the matrix 'A' and a dislocation along the rays which is proportional to $\cos\alpha$. In the case $\cos\alpha$ vanishes, the reflector becomes identical with its caustic, and the image is inversed, as $\cos\alpha$ assumes negative values. However, this would require rather strong deformations. We conclude, therefore, that the caustic pattern formed by the normals remains stable as long as the deformations are of reasonable order.

We note finally that we can reformulate $\cos\alpha\,A$ as

$$(\cos\alpha)A = \frac{1}{2} \begin{bmatrix} 1+\cos2\alpha & -\sin2\alpha \\ \sin2\alpha & 1+\cos2\alpha \end{bmatrix}. \tag{4.71}$$

The caustic formed by the rotated normals can now be written

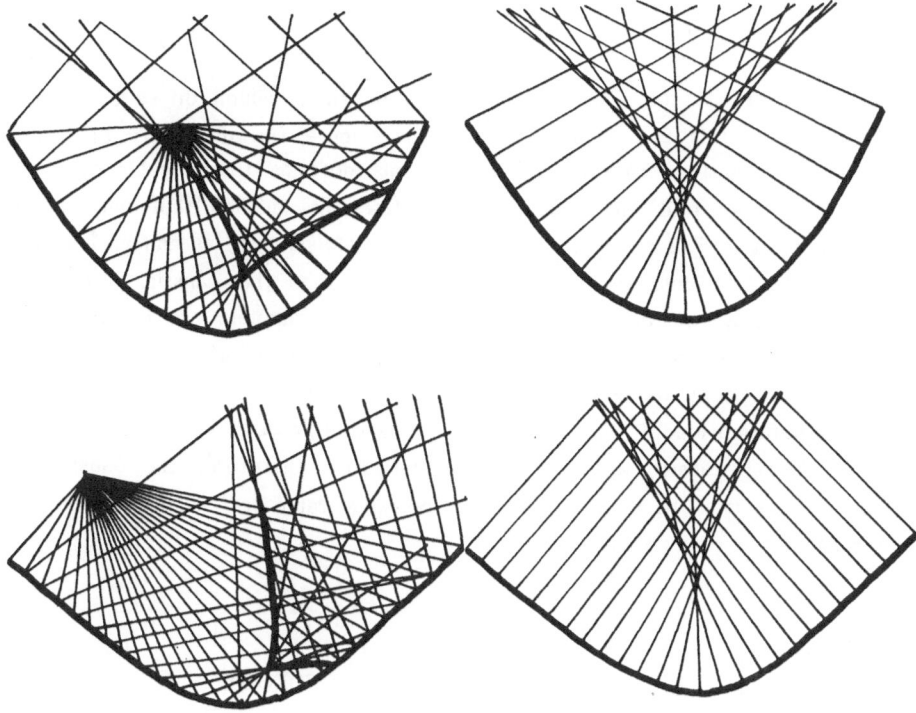

Fig. 4.41: Normal and reflected ray system and caustics at a parabolic and hyperbolic reflector. A point source does not change the caustic pattern.

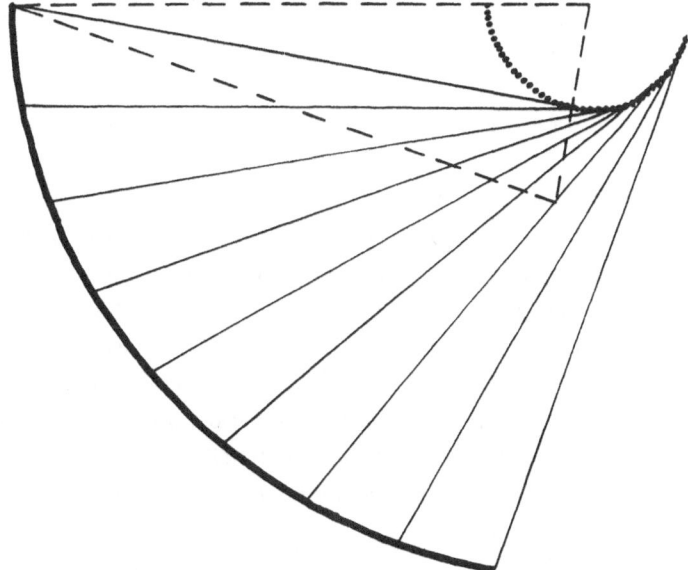

Fig. 4.42: Rotated normals of a circular reflector. The singular point is transformed into a caustic. The rotated rays can be constructed as average of the normals and rays rotated twice the original angle (but which still have the length of the normals).

$$\vec{r}_r = \vec{r} - f(r,r') \frac{1}{2} \left[\vec{n} + \begin{pmatrix} \cos 2\alpha & -\sin 2\alpha \\ \sin 2\alpha & \cos 2\alpha \end{pmatrix} \vec{n} \right] \tag{4.72}$$

The resulting pattern is the average of two vector fields which differ by a rotation. Fig. 4.42 elucidates this point and illustrates once more the instability of a circular reflector. Even a constant rotation of the normals deforms the singular point of the circle into a circular fold line. Fig. 4.42 elucidates in addition that only a rotation with angles larger than $\pi/2$ can really change the caustic pattern, as can also be inferred from equation (4.73). Thus, we can finally conclude that the caustic pattern of the normals of a surface remains stable even if reflected wavelets are received because the rotation of the normal at a circular reflector is equivalent to a deformation of this reflector. This point is especially elucidated by equation (4.73), which states that a monotonous deformation of the reflection angle within reasonable bounds cannot really change the original caustic and the related patterns. Therefore, the caustic formed by the normal rays must be accepted as a structurally stable pattern, even under reasonable disturbances.

4.3 "PARALLEL SYSTEMS" IN GEOLOGY

Structural geology describes and analyzes the "geometry" of deformed rocks. The procedure is mainly geometrical, and the relations to the physical processes are established by "classification procedures" (GZOVSKY et al., 1973). The base for these relationships is developed from various physical, experimental and numerical methods for which a wide variety of mathematical methods has been used (BAYLY, 1974; MATTHEWS et al., 1971; JOHNSON & POLLARD, 1973; BEHZADI & DUBEY, 1980; COBBOLD et al., 1971; DIETRICH, 1970; FLETCHER, 1979; SMITH, 1975 to give a few examples). Most of the geological structures are the result of complex strain fields. These strain fields, in general, are not the result of similar complex global fields of forces, but the complex and inhomogeneous strain field results from the variable elastic, plastic, and viscous behavior of rocks, i.e. from their primary inhomogeneities. The deformations of rocks can be very large, and then they are outside of the scope of classical differential calculus. This is especially true if transitions from elastic to plastic and viscous behavior occur, if the boundary conditions are not known -- in general they are not -- and if dislocations of material by solution and recrystallization play an important role during the deformation process (STEPHANSON, 1974; TRURNIT & AMSTUTZ, 1979). A classical approach to study deformations of rocks, therefore, is the geometrical analysis. A common way is to apply the methods of finite strain analysis (RAMSAY, 1967; JAEGER, 1969; HOBBS, 1971) to regions for which a nearly homogeneous strain can be assumed. Basically, this type of analysis is the study of some special mappings, and some of them will be briefly discussed here.

4.3.1 Some Examples of Parallel Systems

Much of the previous discussion focussed on systems of quasi-parallel layers, which posses a formal geometrical similarity with deformed structures in geology. Considering a three-dimensional space such a parallel system can be written

$$F(u,v;t) = x(u,v) + tN(u,v) \tag{4.73}$$

where $N(u,v) = (N_x, N_y, N_z)$, the unit normal vector at $x(u,v)$.

If $F(u,v;0)$ is everywhere differentiable, then the Jakobian determinant of such a system is nowhere zero (DoCARMO, 1976):

$$\det \; J(F) = \left| \frac{\partial fi}{\partial (u,v,t)} \right| = \mathbf{I} \; (F_u) \; (F_v) \; (F_t) \; \mathbf{I} = \mathbf{I} \; x_u \wedge x_v \; \mathbf{I} \neq 0 \tag{4.74}$$

where F_u etc. are column vectors of the Jacobian matrix (see section 4.3.4 for details).

Equation (4.74) shows that there exists a tubular neighborhood to the surface $x(u,v)$ which is uniquely defined. Given a solution for a surface $x(u,v)$ under certain conditions, we can extend this solution into a small but finite neighborhood of $x(u,v)$. In a conceptual sense this secures that the solutions can be applied to a real physical system where a surface is always of finite thickness. Assume equation (4.73) is applicable as a linear first order approximation, then we immediately get an estimate of the maximal local extent of the tubular neighborhood, i.e. the area into which we may extend the solution. This area is bounded by the 'focal surfaces'

$$x_1(u,v) = x(u,v) + \rho_1 N(u,v) \tag{4.75}$$
$$x_2(u,v) = x(u,v) + \rho_2 N(u,v)$$

where ρ_1, ρ_2 are the principal curvatures of the surface (cf. DoCARMO, 1976; GUGGENHEIMER, 1977).

The assumption of parallel layers has a long tradition in the reconstruction of folds in tectonics (e.g. HILLS, 1963 for an overview). The reconstruction of folds from surface measurements is illustrated in Fig. 4.43 after an example of GILL (1953). A point, obvious from Fig. 4.43, is that the fold cannot be extended continuously into depth, as several segments vanish along the 'caustic of the normal rays' as discussed in the previous section; and it has been assumed (e.g. BUSK, 1956) that these lines (or sufaces) evolve into faults. Of course, equation (4.74) is only a first order approximation; however, rather similar arguments hold for the 'normal variation' of a surface $x(u,v)$ which can be written (DoCARMO, 1976)

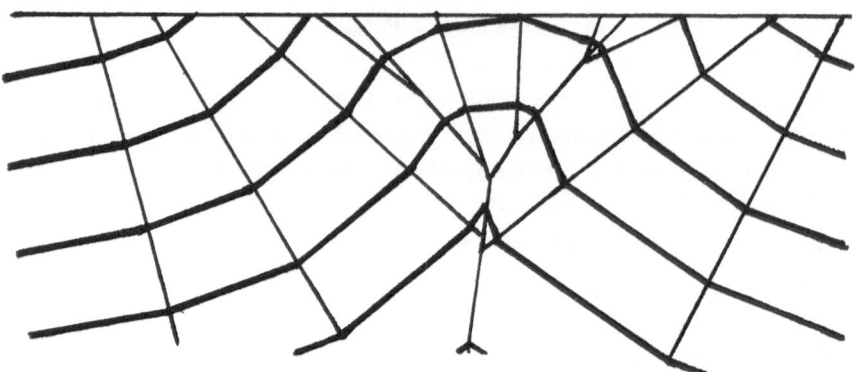

Fig. 4.43: Reconstruction of parallel folds from surface data. Modified after GILL (1953).

$$F(u,v;t) = x(u,v) + th(u,v)N(u,v) \qquad (4.76)$$

where $h(u,v)$ is some scalar variable.

Such a formulation provides us with the possibility to adapt some conditions which have to be satisfied by $h(u,v)$, and equation (4.76) can be considered as a variational problem, or we may consider equation (4.76) as the disturbed linear problem described by equation (4.73).

Parallel systems are encountered in various areas. With respect to geology, an important one is the concept of slip-lines in the theory of perfect plasticity, which is closely related to evolutes and involutes as stated by Hencky's and Prandtl's theorems (LING, 1973):

HENCKY's theorem: The angle formed by the tangents of two fixed shear lines of one family at their points of intersection with a shear line of the second family does not depend on the choice of the intersecting shear line of the second family.

PRANDTL's theorem: Along a fixed shear line of one family, the centers of curvature of the shear lines of the other family form an involute of the fixed shear line.

Given one non-linear shear line, the 'linear system' of normals and involutes provides a first approximation for the slip-line field. The most simple cases are orthogonally intersecting straight lines and 'centered fans' of circular arcs, which provide reasonable first order approximations of plastic deformation (e.g. LING, 1973). Fig. 4.44 illustrates Prandtl's solution for slip-lines below a strip load. A more general solution consists of

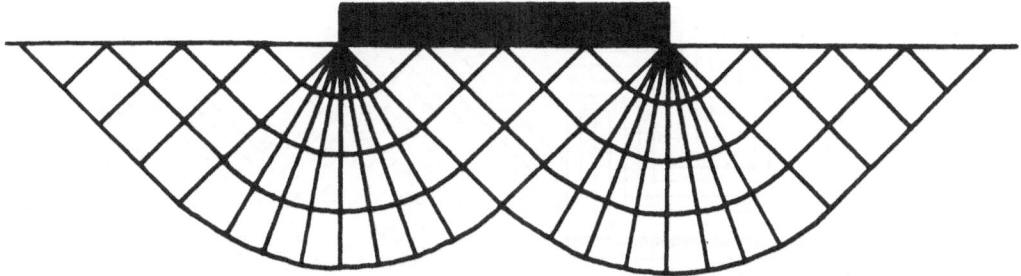

<u>Fig. 4.44:</u> Prandtl's solution for slip-lines below a strip load above a homogeneous halfspace.

centered arcs of logarithmic spirals: Consider $x(u,v)$ a generalized logarithmic spiral, as discussed in section 3.5.3, and $h(u,v)$ to be proportional to the curvature of the leading spiral $x(u,v)$, then equation (4.76) describes a family of possible solutions, from which we have to choose the locally valid one which then can be extended to neighboring areas by connecting local solutions along the straight characteristics.

ODE (1960) applied the slip-line theory to the formation of faults in sand and clay under the conditions of plane strain. By a similar attempt, also more geometrically, FREUND (1974) studied the termination of transcurrent faults by splaying; from his analysis the curvature of transcurrent faults can be related to the formation of an evolute of a fan of faults. Evolutes, as lines (or surfaces) of discontinuity, occur further under unidirectional glide in solid crystals (e.g. KLEMANN, 1983).

4.3.2 Similar and Parallel Folds

Concerning geologically 'shallow' deformations (without phase transitions) HOEPPE-NER (1978) found from experiments that most folds can be traced back to the following types:

        1) similar folds

        2) parallel folds

            a) concentric folds

            b) box folds.

Parallel folds occur usually near the free surface or near shear planes while else-where the more energy consuming similar folds develop. The differences between the

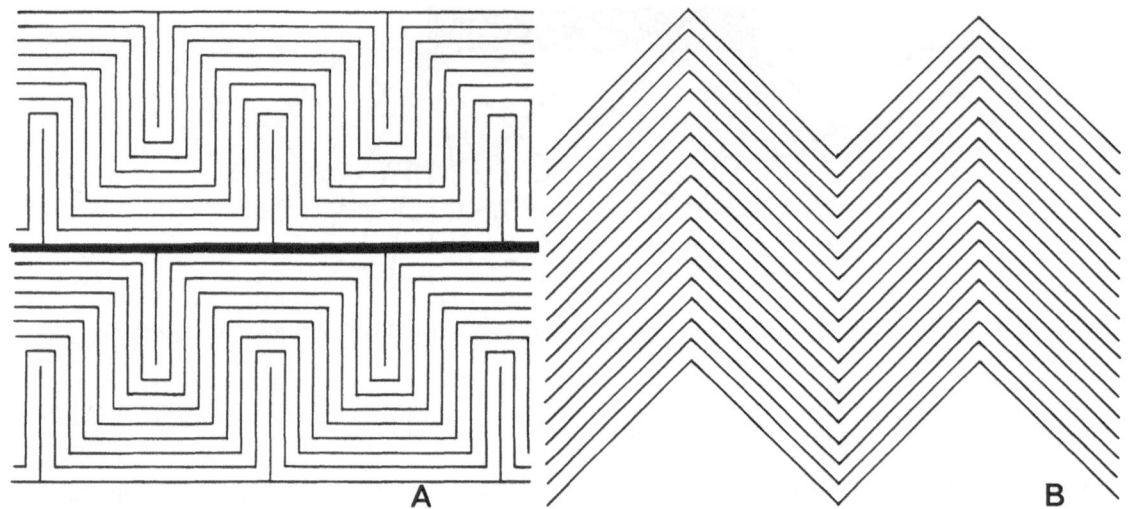

Fig. 4.45: A) Ideal parallel folds (kinks or box folds) and B) ideal similar folds (chevron folds).

two types of folds are schematically illustrated in Fig. 4.45. Parallel folds are of finite depth range, i.e. they resemble the parallel wave fields discussed in the last section. Similar folds in contrary continue (ideally) infinitely. The strain in folded layered systems has extensively be studied by HOBBS (1971), here we consider only volume preserving systems. Similar folds with constant divergence are described by maps of the form

$$X = ax + f(y,z) \qquad (4.77)$$
$$Y = by + h(z)$$
$$Z = cz$$

where a,b,c: constants; f,g,h: arbitrary functions.

The Jacobian determinant

$$\det J = \left| \frac{\partial f_i}{\partial x_j} \right| = abc$$

is constant and by choosing a,b,c in ratios such that abc=1, the deformation described by equation (4.77) is volume conserving, locally and globally. If we consider cylindrical folds, equation (4.77) reduces to a two-dimensional system and describes a two-dimensional dislocation field as illustrated in Fig. 4.46. A special property of this case is that the principal strains are identical along every straight 'shear line' of the dislocation field. As these parallel dislocation lines never intersect, the fold extends ideally into infinite depth, and laterally the local fold pattern can easily be continued if we connect local solutions along a straight dislocation line (cf. Fig. 4.46c). Of course, along such lines the solution is discontinuous with respect to the curvature, a discontinuity which occurs in sinusoidal systems at the inflection points. JOHNSON & ELLEN (1973) pointed out that such lines of discontinuities may be of some value in the analysis of folds

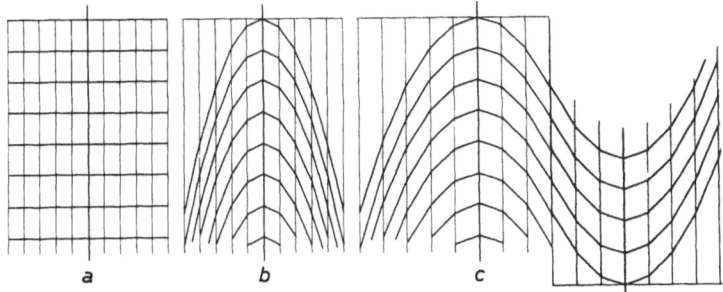

Fig. 4.46: Deformation of a homogeneous half-space (a) into similar folds (b,c). A local solution (c) can be extended laterally by continuation along a slip-line.

and compared them with 'characteristics' as they occur in the slip-line theory of plasticity.

The possibility to continue local solutions laterally is common for both fold types. In the case of parallel cylindrical folds, a local solution can be continued along any 'normal ray' as illustrated by Fig. 4.47. However, for a parallel system there exists no solution with constant Jacobian, and thus it cannot describe a deformation which preserves volume locally. On the other hand, we have already seen in section 4.1.3 that it is possible to connect deformed pieces in such a way that the entire volume of the systems is not altered (Fig. 4.47). Concerning global volume changes, similar and parallel folds provide comparable solutions. Parallel folds are best considered as laminated systems which allow the laminae to glide one above the other as illustrated in Fig. 4.47. Within

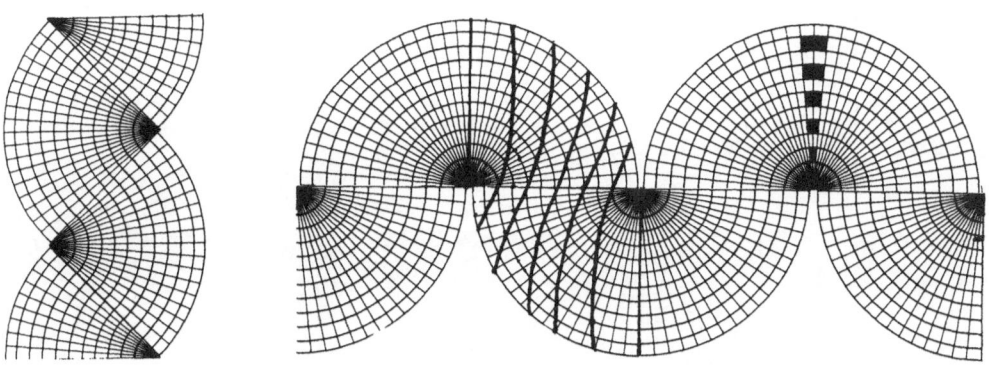

Fig. 4.47: Ideal (concentric) folds laterally continued along 'rays' or 'lines of discontinuity'. Heavy lines indicate intervals of equal length for the layers. Black grid elements: deformed 'volume elements' of originally rectangular grid elements.

204

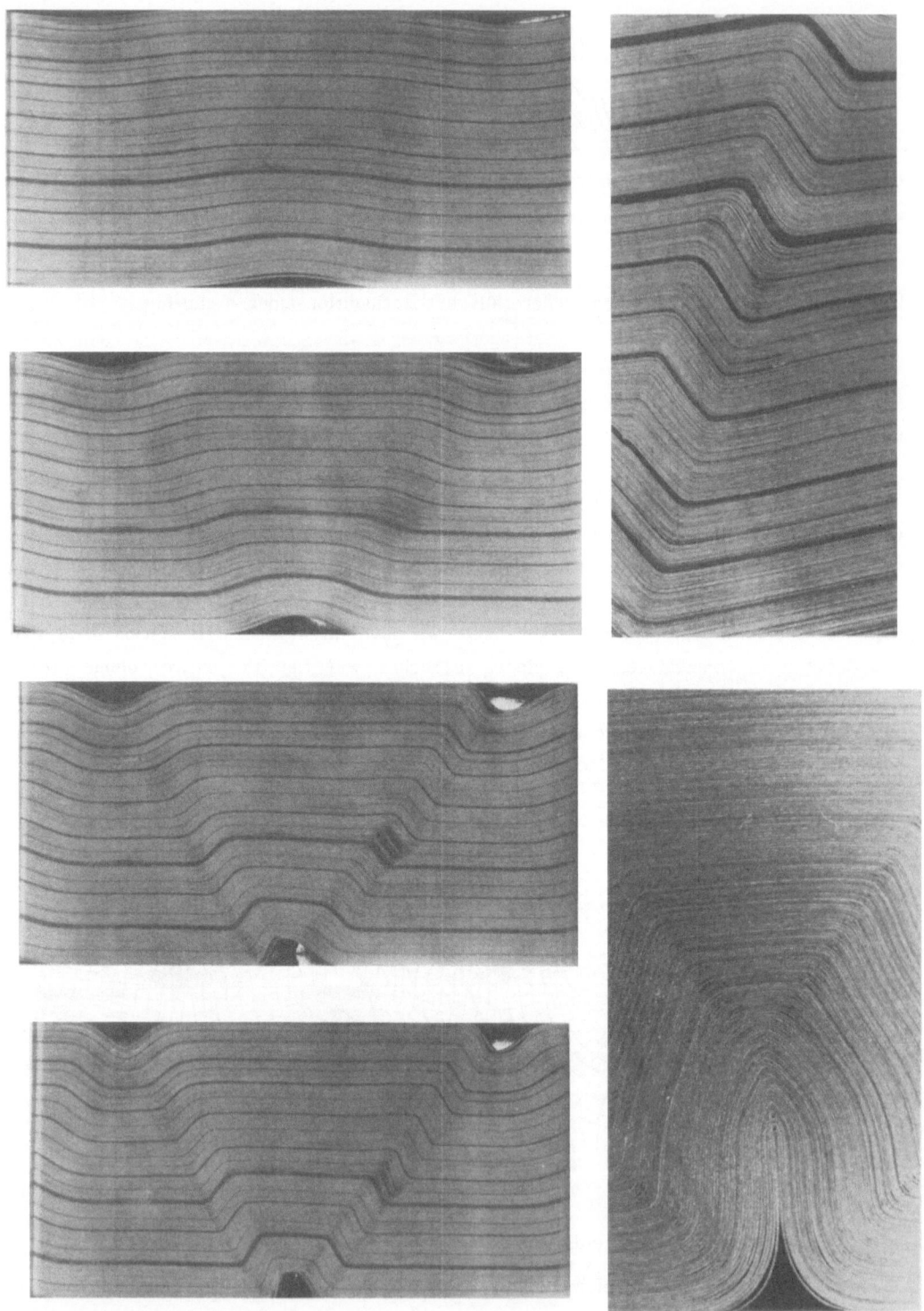

<u>Fig. 4.48:</u> Buckling of a card deck under lateral stress confirmed by vertical plates.
Right: details of kink formation.

laminae the ideal model allows only for membrane stresses, a situation sometimes applicable to deformations in liquid crystals (KLEMAN, 1983). The deformations between subsequent layers then are proportional to the change of surface elements (rather than volume elements):

let $\quad F(u;t) = x(u) + zN(u),$ $\qquad\qquad\qquad\qquad$ (4,78)

then $ds/dt = (1-zk)$

where k: the local curvature of the leading curve,

and the deformation is simply proportional to the curvature of the surface element. We find that the surface elements vanish along the evolute of the normals or at the focal surface, which we, therefore, can expect to be part of the shear surface separating successive parallel folds.

The two fold types are both idealized systems, and experimentally transitions occur between the two types. JOHNSON & HONEA (1975) concluded from their multilayer experiments that the common assumption that one can estimate depth of folding by the 'ray method' is of limited value. Fig. 4.48 illustrates some phases of multilayer folding under two-axial stress. The transition from parallel to similar folds is again a continuation problem. Assume that the solution is known at the free surface of a half space and that this solution is given by a box fold: The range of the parallel fold solution is of limited depth, and it is bounded by a cuspoid focal line (Fig. 4.49). We are interested to extend the disturbance into depth and require that the fold lines are continuous along the focal line. To continue the disturbance we project the focal line into depth, i.e. assume the focal line is given by a function g(x,y) = 0, then we consider the family

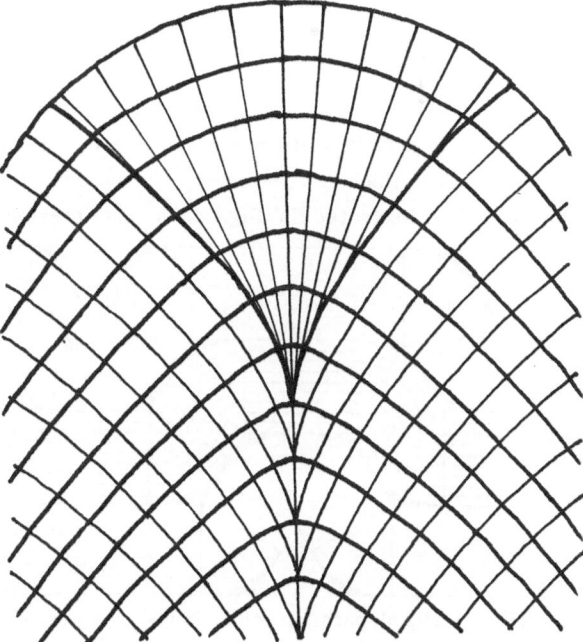

Fig. 4.49: Continuation of parallel folds into similar folds by propagating the cusp discontinuity into depth.

of functions

$$g(x,y) = c. \tag{4.79}$$

The fold lines of the parallel system intersect the focal line perpendicular as was discussed in terms of wave fronts. The extended fold lines, therefore, have also to intersect the original focal line by right angles. A possible continuation, therefore, are the orthogonal trajectories of the family $g(x,y) = c$ which are found by solving the differential equation

$$-g_y + g_x y' = 0. \tag{4.80}$$

In the case of a cuspoid focal line, $x^2 - (y-c)^3 = 0$ or $y - x^{2/3} = c$, the orthogonal trajectories are the family of functions

$$y = (8/9)x^{4/3} + c \tag{4.81}$$

which clearly provide a set of similar folds (Fig. 4.49). Usually the solution will be bounded to a strip of finite length, however, the strip can be continued laterally as discussed previously, and if we consider a layer of finite thickness, this continuation can be adjusted to preserve volume globally. Clearly, the discussed models are only first order approximations which, however, allow graphical analysis of even complicated large scale deformations and which capture some essential qualitative properties of experiments.

4.3.3 Bending at Fold Hinges

The previous discussion focussed on systems composed of layers of vanishing thickness, or of negligible thickness with respect to the entire system. Concerning a compact layer of finite thickness one has to consider the deformations near the fold hinge, as schematically illustrated in Fig. 4.50. The previously discussed linear approach of paral-

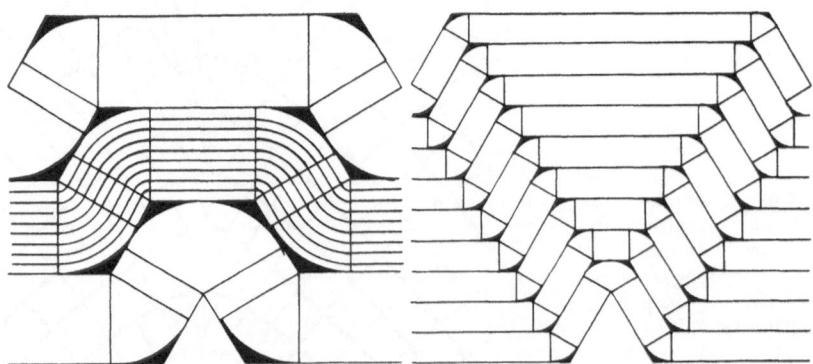

Fig. 4.50: Idealized parallel folds (kinks) composed of layers of different finite thickness.

lel layer reveals Bernoulli's theorem, i.e. undeformed cross-sections in the deformed state. A more realistic model provides St. Vernant's solution for bending of a bar by couples. The deformed state is described by the map

$$X = x(1 + \frac{c}{E} z)$$ (4.82)

$$Y = y(1 - \frac{\sigma c}{E} z)$$

$$Z = z + \frac{c}{2E} (\sigma(y^2-z^2) - x^2)$$

where c: strength of couples; E: Young's modulus, σ : Poisson's ratio and $c/E = R^{-1}$; R: radius of curvature (see e.g. BUDO, 1974; LOVE, 1944).

In engineering the usual procedure is to study the deformation of an object under specific stress configuration. In geology we usually know little about the original stress field. Therefore, it is worthwhile to work with models, and the question is not mainly how the object deforms within a certain stress field but how far the model is applicable. One question, which can be pushed forward by mathematical analysis, is how the various parameters interact and whether the solution is bounded to some region, i.e. concerning the bending model we are interested if the thickness of the bar is unlimited.

The limits of the solution are given by the condition that the Jacobian of St. Vernant's map vanishes; however, in this case we can simplify the analysis by reducing the map to a standard catastrophe on Thom's list. If we slide the bar along the line y=0, i.e. by a vertical plane along the long axis (z), equation (4.82) simplifies to

$$-Z = \frac{c}{2E}(\sigma z^2 + x^2) + z$$ (4.83)

$$X = \frac{c}{E} xz + x,$$

and by means of the transformation $\sigma z^2 \to z^2$ this equation simplifies to the standard form

$$Z^* = z^2 + x^2 + \frac{2E}{c\sqrt{\sigma}} z$$ (4.84)

$$X^* = 2xy + \frac{2E}{c}x.$$

The only assumptions involved are that the couples 'c' do not vanish (we consider only deformed states) and that $\sigma \neq 0$. The singular set of this map is illustrated in Fig. 4.51 by use of the standard form of the hyperbolic umbilic. Fig. 4.52 illustrates a single section (E,c, =constant), and it becomes clear how the solution space is limited by a cusp and a fold line, i.e. even for small deformations of this type the bar cannot exceed a certain thickness. Fig. 4.52a° illustrates the shape of the undeformed area, i.e. the boundaries defined by J=0 (J: Jacobian determinant). By setting J=c one finds lines of

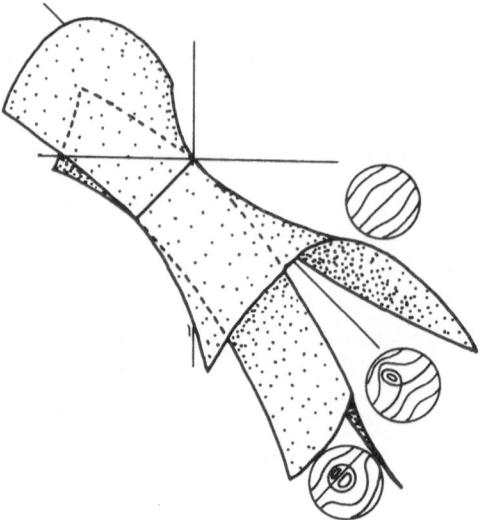

Fig. 4.51: The standard form of Thom's hyperbolic umbilic. Isolines inside circles indicate the local 'potential'.

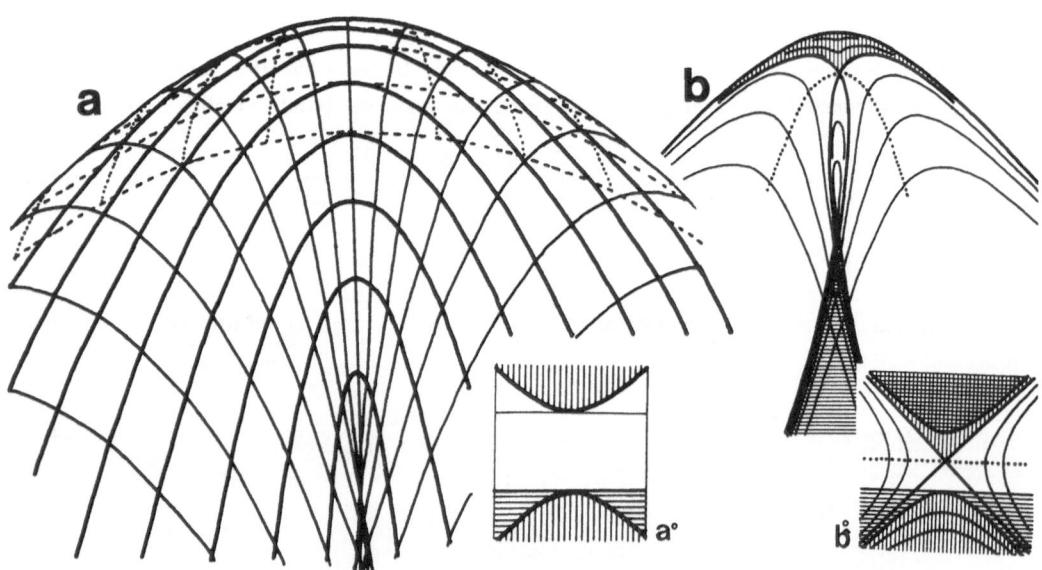

Fig. 4.52: a) The non-local section through a bent bar along its long axis (see text). The critical set (self-intersections of parabolas) corresponds with the section through the hyperbolic umbilic. Parabolas indicate lines which are parallel in the undeformed state. a°) associated undeformed state: Only the blank area can be deformed to the image indicated in (a). b) The same bar with lines of constant values of the Jacobian determinant. b°) the associated undeformed image.

209

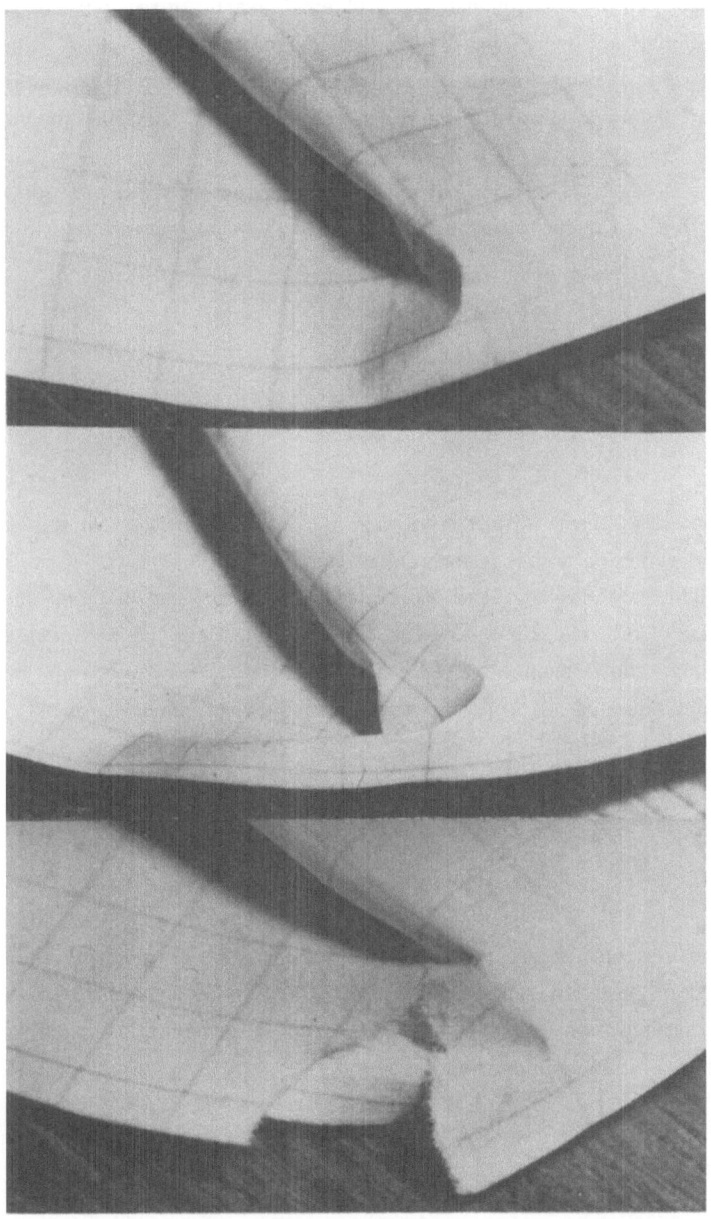

<u>Fig. 4.53:</u> The 'hyperbolic umbilic' as a sheet of paper folded in its plane.

"equal volume change" in the undeformed state (Fig. 4.52b°) and by means of the mapping (4.83 or 4.84) their image in the deformed state (Fig. 4.52b). Further properties will be analyzed in a more general sense in the next sections. The reduction of the original map to a two-dimensional problem, clearly, gives only an idea how the entire system reacts; however, the solution is correct for the plane selected, and it allows to relate the deformation to a rather simple experiment (Fig. 4.53): A sheet of paper 'bended' in its plane illustrates in a rather simple way how the limiting fold line evolves.

To connect this section with the further analysis of parallel systems we observe that equation (4.83) can alternatively be written (using vector notation):

$$
\begin{bmatrix} X \\ Y \\ Z \end{bmatrix} = \begin{bmatrix} x \\ y \\ \frac{c}{2E}(\sigma y^2 - x^2) \end{bmatrix} + \frac{c}{E} z \begin{bmatrix} x \\ -\sigma y \\ E/c \end{bmatrix} - \frac{c}{2E} z^2 \begin{bmatrix} 0 \\ 0 \\ \sigma \end{bmatrix} \tag{4.85}
$$

where the first term is just the description of the "neutral surface" and the second term is the non-normalized normal of the surface elements. Thus, the first two terms on the right side describe the 'normal variation' of the surface, i.e. $h(u,v) = |X_u \wedge X_v|$ in equation (4.76). The final term on the right side can be taken as a non-linear disturbance of the quasi-parallel system. This non-linear term depends only on z such that the bending equation is properly approximated by the quasi-parallel system if z is sufficiently small.

4.3.4 Notation of Strain

Whenever elastic or plastic deformations are considered, the problem is usually formulated in terms of stresses and strains. The procedure is to solve a given problem in terms of dislocations (e.g. LOVE, 1944). The deformed state then is given in the form discussed with similar folds, i.e. by a map

$$
\begin{bmatrix} X_1 \\ X_2 \\ X_3 \end{bmatrix} = \begin{bmatrix} x_1 \\ x_2 \\ x_3 \end{bmatrix} + \begin{bmatrix} \xi_1 \\ \xi_2 \\ \xi_3 \end{bmatrix} \quad ; \qquad \xi_i = f(x_1, x_2, x_3) \tag{4.86}
$$

deformed   undeformed   dislocations

The elements of strain are related to the Jacobian matrix of the dislocations (e.g. LOVE, 1944).

$$J(\underline{\Xi}) = \begin{bmatrix} \dfrac{\partial \xi_1}{\partial x_1} & \dfrac{\partial \xi_1}{\partial x_2} & \dfrac{\partial \xi_1}{\partial x_3} \\[2mm] \dfrac{\partial \xi_2}{\partial x_1} & \dfrac{\partial \xi_2}{\partial x_2} & \dfrac{\partial \xi_2}{\partial x_3} \\[2mm] \dfrac{\partial \xi_3}{\partial x_1} & \dfrac{\partial \xi_3}{\partial x_2} & \dfrac{\partial \xi_3}{\partial x_3} \end{bmatrix} = \left[\left(\dfrac{\partial \Xi}{\partial x_1}\right)\left(\dfrac{\partial \Xi}{\partial x_2}\right)\left(\dfrac{\partial \Xi}{\partial x_3}\right) \right] \tag{4.87}$$

where (Ξ_x) etc. are the column vectors of the Jacobian matrix.

Now, there exists a simple relationship between the Jacobian matrix of the map (4.86), which will be denoted by $J(X)$, and the Jacobian matrix of the dislocations

$$J(X) = J(\underline{\Xi}) + I \tag{4.88}$$

where I: the identity matrix.

If we now consider the more general map

$$\begin{bmatrix} X_1 \\ X_2 \\ X_3 \end{bmatrix} = \begin{bmatrix} u(x_1, x_2, x_3) \\ v(x_1, x_2, x_3) \\ w(x_1, x_2, x_3) \end{bmatrix}, \tag{4.89}$$

and the Jacobian matrix of the dislocations is

$$J(\Xi) = J(X) - I.$$

The matrix of strain elements in the linear theory of elasticity then is given by (e.g. LOVE, 1944; MEANS, 1976)

$$(e_{ij}) = \frac{1}{2}\left\{ \left[\left(\frac{\partial X_i}{\partial x_j}\right) + \left(\frac{\partial X_j}{\partial x_i}\right)\right] - I \right\}. \tag{4.90}$$

In the nonlinear theory of finite strain, however, the matrix of strain elements can be written

$$(\varepsilon_{ij}) = \frac{1}{2}\left\{ \left[\left\langle \left(\frac{\partial X_i}{\partial x_j}\right), \left(\frac{\partial X_j}{\partial x_i}\right)\right\rangle\right] - I \right\} \tag{4.91}$$

where $<\,,\,>$ denotes the scalar product of the column vectors of the Jacobian matrix. We shall need these notations because they simplify the following work.

4.3.5 Generalized Plane Strain in Layered Media

Geological problems are commonly solved under the assumption of plane strain. In this case, the three-dimensional problem simplifies to a two-dimensional one, which causes less computational problems even in the computer. Here we consider the case of generalized plane strain and define it by the condition that $e_{zz}=e_{xz}=e_{yz}=0$. If we apply this to the equations of similar folds (equation 4.77), we find that the state of plane strain is achieved by a map

$$X = ax + f(y)$$
$$Y = by + g(x) \tag{4.92}$$
$$Z = z$$

where not only the strain components involving the z-direction vanish but also the associated dislocations. Special cases of this type have been discussed by HOBBS (1971) in detail. A more interesting situation arises if one considers parallel folds. We take the discussion of the bending of bars as motivation and consider the quasi-parallel system

$$\begin{bmatrix} X \\ Y \\ Z \end{bmatrix} = \begin{bmatrix} x \\ y \\ f(x,y) \end{bmatrix} + z \begin{bmatrix} -f_x \\ -f_x \\ 1 \end{bmatrix} \tag{4.93}$$

or $\quad X = x(u,v) + z|x_u \wedge x_v| N$

and apply the linear theory of strain. The strain elements are found from equation (4.90)

$$e_{xx}=-zf_{xx}; \qquad e_{yy}=-zf_{yy}; \qquad e_{xy}= zf_{xy}; \tag{4.94}$$
$$e_{zz}=e_{xz}=e_{yz}=0;$$

i.e. the quasi-parallel system describes a state of generalized plane strain. Next we consider the perfect parallel system

$$X = x(u,v) + zN(u,v) \tag{4.95}$$

and apply the non-linear theory of finite strain. The Jacobian matrix of this map can be written

$$J(X) = \big((x_u + zN_u), (x_v + zN_v), (N)\big). \tag{4.96}$$

The finite strains are defined by the scalar products of the column vectors of J (equ. 4.91) which simplify if one applies the orthogonality relations $<x_u,N_u> =0; < x_u,N_v> =0$ etc.. The finite strains are

$$\varepsilon_{xx} = \frac{1}{2}(<(x_u+zN_u),(x_u+zN_u)> -1) = \frac{1}{2}(<x_u,x_u> + 2z<N_ux_u> + z^2<N_u,N_u>) \tag{4.97}$$

$$\varepsilon_{yy}=\frac{1}{2}(<(x_v+zN_v),(x_v+zN_v)> -1)= \frac{1}{2}(<x_v,x_v> + 2z<N_vx_v> + z^2<N_v,N_v> -1)$$

$$\varepsilon_{xy} = \frac{1}{2}(\langle (x_u + zN_u),(x_v + N_v)\rangle - 1) = \frac{1}{2}(\langle x_u, x_v \rangle + z(\langle N_u, x_v \rangle + \langle N_v x_u \rangle) + z^2 \langle N_u, N_v \rangle)$$

$$\varepsilon_{zz} = \varepsilon_{xz} = \varepsilon_{yz} = 0.$$

The two equations, thus, are isomorphic with respect to the applied theory of strain, and because the linear theory is an approximation of the non-linear one, equation (4.93) provides an approximation of equation (4.95). Indeed, if we consider only the "neutral surface" X = x(u,v) which is everywhere differentiable, then we can approximate it locally by an explicit function z=f(x,y) in terms of its moving frame. There is another important aspect: If we consider the case x(u,v)=f(u,v), then the focal surfaces, which bound the solution space, are identical for equation (4.93) and equation (4.95) because we can define them as the evolute (surface) of the normal rays, and the only difference between these equations is the magnitude of dislocation along the normals. Thus, even the solutions deviate to some extent in the linear and non-linear model, the critical set of focal domains remains identical.

Both equations studied here describe families of surfaces, and this fact has its expression in the equations for the strains, which can be expressed in terms of the fundamental forms of the "neutral surface". Following DoCARMO (1976) we note that

$$E = \langle x_u, x_u \rangle; \quad G = \langle x_v, x_v \rangle; \quad F = \langle x_u, x_v \rangle$$

$$-e = \langle N_u, x_u \rangle; \quad -g = \langle N_v, x_v \rangle; \quad -2f = \langle N_u, x_v \rangle + \langle N_v, x_u \rangle,$$

and the elements of strain can be written

$$\begin{aligned}
\varepsilon_{xx} &= E - 2ze + z^2 \langle N_u, N_u \rangle \\
\varepsilon_{yy} &= G - 2zg + z^2 \langle N_v, N_v \rangle \\
\varepsilon_{xy} &= F - 2zf + z^2 \langle N_u, N_v \rangle
\end{aligned} \tag{4.98}$$

providing the base for a potential further analysis.

The study of parallel system thus provides us with rather general models for generalized plane strain and with strategies to find more general solutions in terms of finite strain than usually: If we are able to find a solution in terms of the linear theory, the close relations discussed here allow to transfer it to the non-linear theory as far as quasi-parallel systems are concerned. The equations governing bending deviate from the quasi-parallel linear model only by an additional non-linear term. In the finite strain model a rather similar structure is achieved if the variable 'z' is replaced by a function h(z), e.g. h(z) = h(1-z).

The quasi-parallel systems provide a family of functions of potential value for the analysis of large scale deformations, classifiable by the structure of their Jacobian matrix (or the related strains) -- an aspect which links this study with catastrophe theory. The problems, which have been discussed here, are those which can be solved by 'hand methods'. In layered media the physical properties usually alternate (or change gradually). Replacing the scalar function $h(x,y)$ in equation (3.76) by a matrix $H(x,y;z)$ allows to study more complicated systems in terms of iterated maps.

4.4 SUMMARY

The discussed examples are manifold, covering various geological and "applied mathematical" methods. This calls for a systematization of the various forms of instable behavior and pattern formation. There are two aspects: First, we have three major mathematical objects -- surfaces, trajectories, and distance functions -- which, under certain maps, develop discontinuities or become instable in some sense. Secondly, we have the instabilities themselves which cause branching solutions.

A first group of objects, which occured repeatedly throughout the examples, are surfaces which can locally be described as $F(x_1, \dots, x_i)=0$. The interesting deformations occur under the transformation $F=F(x_1, \dots, x_i) + sN(x_1, \dots, x_i)$. The surface transforms like a wave front, at least locally, which is constructed by Huygens' principle. This implies that the family of surfaces can locally be described as $F(x_1, \dots, x_i; s)=0$, and the concept of structural stability can be used to study the geometrical singularities. The evolution of wave fronts is, of course, the typical example although the discontinuities, which arise in the two-dimensional projection of three-dimensional objects, provide still more geometrical examples. In addition, we can locate here the probabilistic hull of the ontogenetic 'morphospace' of chapter 2, and if we allow for more general dislocations, then the whole example of the ontogenetic 'morphospace' belongs to this class of problems. Clearly, the "pre-computer" analysis of parallel folds and the slip-line theory are closely related in geometrical terms although the physical parameters are quite different.

The second group of objects were trajectories -- ontogenetic traces in chapter 2, rays and slip-lines in chapter 4. Along these trajectories a dynamics can be established -- the spreading of a wave front, the ontogenetic development etc. -- by an equation like $ds/dt=f(s)$ (s:arc length). But the trajectories or rays themselves are static objects like the previously discussed families of surfaces, and they are described by the identical equations up to a normalization factor of the evolution parameter. If one has especially a gradient system, then elementary catastrophe theory provides a classification of the interesting geometrical singularities. However, the gradient restriction is not so essential because elementary catastrophe theory is a local theory, i.e. it is sufficient if we can transform such a system locally, near a critical point, into a gradient system. The interesting singularities are usually different from the 'involutes of rays'; however, they are related as discussed in terms of seismic reflection.

A further group of examples can be collected under the term "distance functions", and this is the most comprehensive group. Again, we can easily return to elementary catastrophe theory. The wave fronts on a given set of rays have to satisfy the distance function $r=|(x,y)-(x_o,y_o)|$. Similar formulations are possible for the slip-line theory and elasticity theory using the concept of strain and stress potentials. It is commonly the approach via a distance or energy function, which allows to study problems more deeply in topological terms. In principle, the problems encountered in surface reconstruction and the convex hulls of point sets (chapter 2, Honda trees) belong to this group.

If we now take the example of cluster analysis, the situation is different. What we are doing in this case is essentially that we map a n-dimensional space onto a one--dimensional one, the space of distances. This allows to represent clustering results from a n-dimensional space as a binary tree. The binary tree is nothing than the graphical representation of the pair (X,d) where 'X' is a one-dimensional point set and 'd' is a relation between the points which defines an order. In the case of cluster trees, the problem is that the distance 'd' does not imply an ordering sequence and that the map $R^n \rightarrow R$ can be multi-valued. This returns the problem to singularity theory as encountered with the two-dimensional images of three-dimensional objects.

What does hold together all examples, is that we can describe them all as maps and that the observed instabilities are related to singularities of these maps whereby the identical geometrical singularity may have different physical meaning concerning different objects and may be different with respect to any chosen subspace, i.e. in spatial and spatio-temporal coordinates. Maps in their most general sense are not very specific. Insofar, the previous discussion provides examples of more specific maps, which are of some interest in geological applications.

Concerning the bifurcation of solutions the preceding discussion remains within kinematic models. A bifurcation point is defined as the value of an evolution parameter at which the local topology changes. Thus, the cusp catastrophe is the bifurcation set of the swallowtail, it separates the area with continuous wave fronts from that one with selfintersecting wave fronts which, of course, are of different topological type, and in a similar sense the transition from parallel to similar folds can be interpreted, etc.. The catastrophe approach "implies that a non-bifurcation point is one at which the topology does not change, i.e. a point at which the system is structurally stable. ... bifurcation is seen as a loss of (structural) stability of the system, rather than the

216

loss of stability of a particular solution" (STEWART, 1982). To avoid nomenclature conflicts, these topological bifurcations are called catastrophes. Because of their wide meaning and their topological objectives they appear well suited for new approaches of geometrical reasoning in geology.

ELEMENTARY CATASTROPHES

POTENTIALS AND CRITICAL SETS

fold
$$V(x) = \frac{1}{3}x^3 + ux$$

cusp
$$V(x) = \frac{1}{4}x^4 + \frac{1}{2}ux^2 + vx$$

swallowtail
$$V(x) = \frac{1}{5}x^5 + \frac{1}{3}ux^3 + \frac{1}{2}vx^2 + wx$$

hyperbolic umbilic
$$V(x,y) = x^3 + y^3 + wxy - ux - vy$$

elliptic umbilic
$$V(x,y) = x^3 - 3xy^2 + w(x^2 + y^2) - ux - vy$$

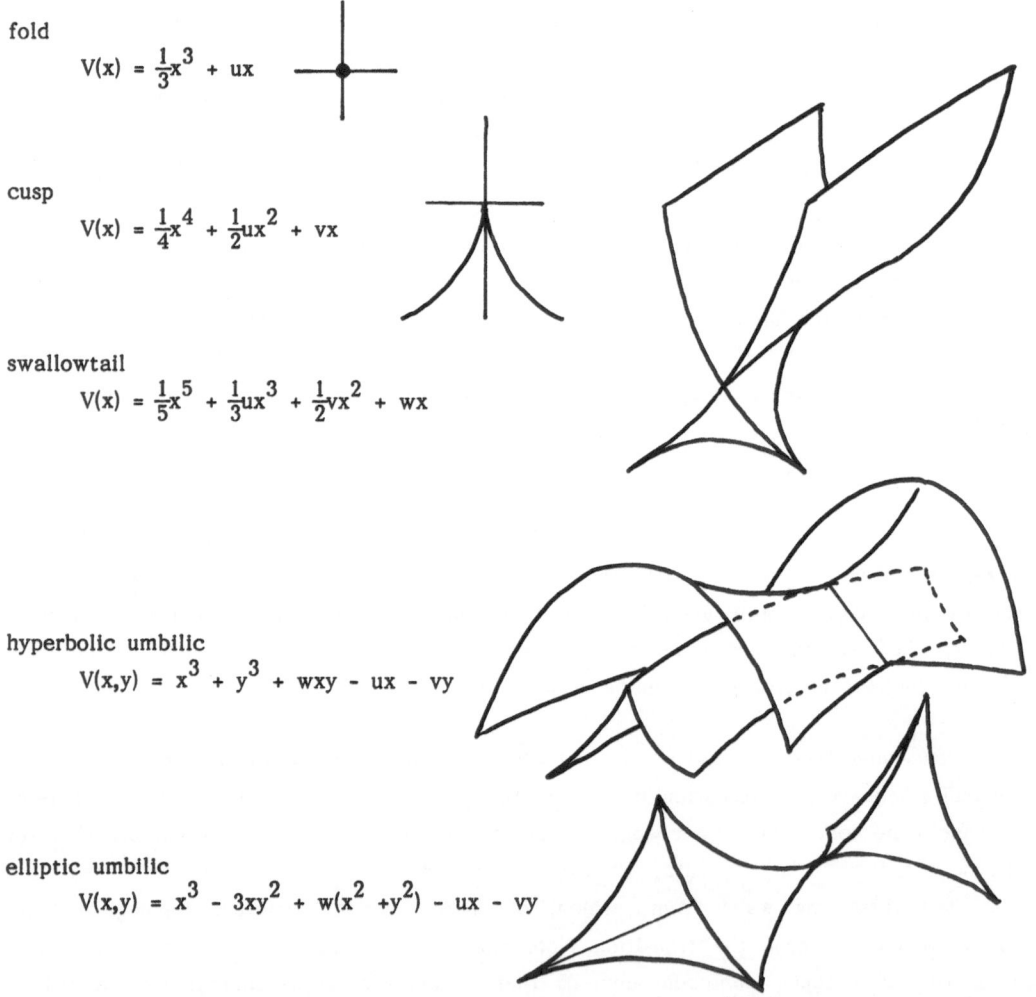

These are only those elementary catastrophes which are easily graphed. For a detailed list see one of the textbooks.

REFERENCES

Adler, R.E. 1970: Elektronische Datenverarbeitung in der modernen Tektonik. Clausth. Tekt. Hefte, 10:25-47.

Adler, R.E. et al. 1968: Elektronische Datenverarbeitung in der Tektonik. Clausth. Tekt. Hefte, 8.

Alagić, S. & Arbib, M.A. 1978: The design of well-structured and correct programs. (Springer), New York.

Altheimer, E., Bayer, U., Ott, R. 1982: The mapping package PEMAP. In: Seilacher, A. et al.eds.:Paleocology. N. Jb. Geol. Pal., Abh., 164:17-23.

Andronov, A.A., Vit, A.A., Khaikin, C.E. 1966: Theory of oscillators. (Pergamon) Oxford.

Arnold, V.I. 1984: Catastrophe theory. (Springer) Heidelberg.

Ashby, W.R. 1956: An introduction to cybernetics.
--   1974: Einführung in die Kybernetik. Frankfurt.

Ballentyne, C.K. & Cornish, R. 1979: Use of the Chi-square test for the analysis of orientation data. Geol. Mag.

Barker, S.B., Cumming, G., Horsfield, K. 1973: Quantitative morphometry of the branching structure of trees. J. theor. Biol., 40:33-43.

Bayer, U. 1969: Die Gattung Hyperlioceras BUCKMAN (Ammonoidea, Graphoceratidae) aus dem Unter-Bajocium (discites-Schichten), insbesondere vom Wutachtal (Südbaden).

   Jber. u. Mitt. oberrhein. geol. Ver., N.F. 51:31-70.

-- 1970: Anomalien bei Ammoniten des Aaleniums und Bajociums und ihre Beziehung zur Lebensweise. N. Jb. Geol. Paläont. Abh., 135:19-41.

-- 1972: Zur Ontogenie und Variabilität des jurassischen Ammoniten Leioceras opalinum. N. Jb. Geol. Paläont. Abh., 140:306-327.

-- 1977: Cephalopoden-Septen. Teil 1: Konstruktionsmorphologie des Ammoniten-Septums. N. Jb. Geol. Paläont. Abh., 154:290-364.

-- 1977: Cephalopoden Septen. Teil 2: Regelmechanismen im Gehäuse- und Septenbau. N. Jb. Geol. Paläont. Abh., 155:162-215.

-- 1978: Morphogenetic programs, instabilities and evolution -- a theoretical study. N. Jb. Geol. Paläont. Abh., 156:226-261.

-- 1978: Models in morphogenesis. N. Jb. Geol. Paleont. Abh., 157:57-70.

-- 1978: Modelle instabiler Kompaktionsverläufe unter Sedimentation. Geolog. Rundschau, 67:980-990.

-- 1978: Finite computations in compaction theory. In: F. Westphal & A. Seilacher eds.: Palökologie. N. Jb. Geol. Paläont. Abh., 157:176-185.

-- 1982: Data storage and data processing in paleoecology. In: A. Seilacher et al. eds.: Studies in Palecology. N. Jb. Geol. Paläont. Abh., 164:23-25.

-- 1982: Cluster- and classification-strategies in paleoecology. In: Seilacher et al. eds.: Studies in palecology, N. Jb. Geol. Paläont. Abh., 164:12-17.

-- 1982:Zur Auswertung von Richtungsdaten. In: Seilacher et al. eds.: Studies in Palecology, N. Jb. Geol. Paläont. Abh., 164:23-25.

-- 1982: Wahrscheinlichkeitsmaße auf Profilen. In: Seilacher et al. eds.: Studies in Palecology, N. Jb. Geol. Paläont. Abh., 164:26-29.

-- 1983: "Physical properties" in: Ludwig, W.J., Krasheninikov, V.A., et al.: Initial Rep. Deep Sea Drilling Project, Vol. LXXI, (U.S. Gov. Print. Off.) Washington.

Bayer, U. 1983: The influence of sediment composition on physical properties inter-relationships. In: Ludwig, W.J., Krasheninikov, V.A. et al.: Initial Rep. Deep Sea Drilling Project, Vol. LXXI (U.S. Gov. Print. Office) Wahington.

Bayer, U. & McGhee, G.R. 1984: Iterative evolution of middle Jurassic ammonite faunas. Lethaia, 17:1-16.

Bayer, U. & Seilacher, A., eds.: Sedimentary and evolutionary cycles. (Springer) Lect. Notes Earth Sci.,1. 1985.

Bayly, M.B. 1974: An energy calculation concerning the roundness of folds. Tectono-physics, 24:291-316.

Behzadi, H. & Dubey, A.K. 1980: Variation of interlayer slip in space and time during flexural folding. Journ. struct. Geol., 2:453-457.

Ben-Menahem, A. & Singh, S.J. 1981: Seismic waves and sources. (Springer) New York.

Betz, A. 1948: Konforme Abbildungen. Berlin.

Biju-Duval, B. et al. 1974: Geology of the Mediterranean Sea basins. In: Burk, C.A. & Drake, C.L. eds.: The geology of continental margins. (Springer) New York, 695-721.

Blackith, R.E. & Reyment, R.A. 1971: Multivariate morphometrics. London.

Blind, W. 1976: Die ontogenetische Entwicklung von Nautilus pompilius (LINNE). Paleontographica, A, 153:117-160.

Blum, H. 1973: Biological shape and visual science. J. Theor. Biol. 38:205.

Blum H. & Nagel, R.W. 1977: Shape description using weighted symmetric axis features Proc. I.E.E.E. Comp. Soc. Conf. Pattern Recogn. Image Proc.

Bonner, J.T. 1968: Size changes in development and evolution. J. Paleont. 42:1-15.

Bonyum, D. & Stevens G. 1971: A general purpose computer program to produce geologi-cal stereonet diagrams. In: Cutbill, J.L. ed.: Data processing in biology and geology. (Acad. Press) New York, 165-188.

Bookstein, F.L. 1978: The Measurement of biological shape and shape change. Lect. Notes Biomath. 24 (Springer) Berlin.

Boyce, R.E. 1977: Deep Sea Drilling procedures for shear strength measurements of clayey sediments using modified Wykeham Farrance Laboratory Vane Apparatus. In: Barker, Dalzill et al.: Init. Rep. DSDP, 36 (U.S. Gov. Print. Office) Washington ,837-847.

Buckman, S.S. 1887-1907: Monograph of the ammonites of the Inferior Oolite series. Paleontogr. Soc. London.

Budó, A. 1974: Theoretische Mechanik. (VEB) Berlin.

Burk, C.A. & Drake, C.L. 1974: The geology of continental margins. (Springer) New York.

Busk, H.G. 1956: Earth Flexures. Cambridge.

Callahan, J.J. 1974: Singularities and plane maps. Amer. Math. Monthly, 81:211-240.

Cavendish, J.C. 1974: Automatic triangulation of arbitrary planar domains for the finite element method. Int. J. Numer. Method. in Engineering, 8:679-696.

Chilingarian, G.V. & Wolf, K.H. eds. 1975-1976: Compaction of coarse grained sedi-ments I,II. Developments in Sedimentology, 18A,B, (Elsevier) New York.

Cobbold, P.R., Cosgrove, J.W. Summers, J.M. 1971: Development of internal structures in deformed anisotropic rocks. Tectonophysics, 12:23-53.

Collatz, L. & Wetterling, W. 1971: Optimierungsaufgaben. (Springer) Berlin.

Cope, J.C.W. et al. 1980: A correlation of Jurassic rocks in the British Isles. Part I & II. Geol. Soc. London Spec. Rep. 14/15.

Courant, R. & Hilbert, D. 1968: Methoden der mathematischen Physik. (Springer) Berlin.

Damuth, J. E. 1980: Use of High-Frequency (3.5–12kHz) Echograms in the study of Near-Bottom Sedimentation Processes in the Deep-Sea: A Review. Marine Geology, 38:51–75.

Dangelmayr, G. & Armbruster, D. 1983: Singularities in phonon focusing. Preprint Inst. f. Informationsverarb. Tübingen.

Dangelmayr, G. & Güttinger, W. 1980: Remote sensing in terms of singularity theory. Mskr. Inst. f. Informationsverarbeitung Tübingen.

Dangelmayr, G. & Güttinger, W. 1982: Topological approach to remote sensing. Geophys. J. R. Astr. Soc., in press.

Davis, J.C. 1973: Statistics and data analysis in geology. (J.Wiley) New York.

Davis, J.G. & McCullagh, M.J. eds. 1975: Analysis of spatial data. (J. Wiley) London

DeBoor, C. 1978: A practical guide to splines. (Springer) Appl. math. Sci. 27, New York.

Denert, E. & Frank, R. 1977: Datenstrukturen. (BI) Zürich.

Desai, Ch.S. & Christian J.T. 1977: Numerical methods in geotechnical engineering. (McGraw-Hill) New York.

Diday, E. & Simon, J.C. 1980: Clustering analysis. In: Fu, K.S. ed.: Digital pattern recognition. 2nd ed. (Springer) Berlin, 47–94.

Dietrich, J.H. 1970: Computer experiments on mechanics of finite amplitude folds. Canadian J. Earth Sci., 7:467–476.

DoCarmo, M.P. 1976: Differential geometry of curves and surfaces. (Prentice-Hall) Engelwood Cliffs.

Donath, F.A. 1969: Experimental study of kink-band development in the Martinsburg slate. Geol. Surv. Canada, 68–52:255–288.

Driver, E.S. & Pardo, G. 1974: Seismic traverse across the Gabon continental margin. In: Burk, C.A. & Drake C.L. eds.: The geology of continental margins.(Springer) New York, 293–311.

Duffy, M.R., Britton, N.F., Murray, J.D. 1980: Spiral wave solutions of practical reaction diffusion systems. SIAM J. Appl. Math., 39:8–13.

Efron, B. 1965: The convex hull of a random set of points. Biometrika, 52:331–343.

Einsele, G. 1977: Range, velocity, and material flux of compaction flow in growing sedimentary sequences. Sedimentology, 24:639–655.

Einsele, G. 1985: Response of sediments to sea-level changes in differing subsiding storm-dominated marginal and epeiric basins. In: Bayer & Seilacher eds.: Sedimentary and evolutionary cycles. (Springer) Heidelberg.

Einsele, G. & Seilacher, A. eds. 1982: Cyclic and event stratification. (Springer) Berlin.

Eldredge, N. & Gould, S.J. 1972: Punctuated equilibria: An alternative to phyletic gradualism. In: Schopf, T.J.M. ed.: Models in paleobiology. (Freeman), San Francisco, 82–11.

Embley, R.W. 1980: The role of mass transport in the distribution and character of deep-ocean sediments with special reference to the North Atlantic. Marine Geology, 38:23–50.

Fisher, J.B. & Honda, H. 1977: Computer simulation of branching pattern and geometry in Terminalia (Combretaceae), a tropical tree. Bot. Gaz. 138, 377–384.

Fisz, M. 1976: Wahrscheinlichkeitsrechnung und mathematische Statistik. (VEB) Berlin

Fletcher, R.C. 1979: The shape of single-layer folds at small but finite amplitude. Tectonophysics, 60:77-87.

Flood, R.D. 1980: Deep-Sea sedimentary morphology, modelling and interpretation of echo-sounding profiles. Marine Geology, 38:77-92.

Fox, T. 1975: Some practical aspects of time series analysis. In: McCammon R.B. ed.: Concepts in Geostatistics. (Springer), Berlin, 70-89.

Freeman, H. & Pieroni, G.G. eds. 1980: Map data processing. (Acad. Press) New York.

Freund, R. 1974: Kinematics of transform and transcurrent faults. Tectonophysics, 21:93-134.

Fu, K.S. ed. 1980: Digital pattern recognition. (Springer) Berlin.

Gebelein, H. 1951: Anwendung gleitender Durchschnitte zur Herausarbeitung von Trend-linien und Häufigkeitsverteilungen. Mitt. Bl. math. Statistik, 3:45-68.

Gill, W.D. 1953: Construction of geological sections of folds with steep-limb atten-uation. Bull. Amer. Assoc. Petrol. Geol., 37:2389-2406.

Gould, S.J. 1966: Allometry and size in ontogeny and phylogeny. Biol. Rev., 41:587-640.

-- 1971: Geometric similarity in allometric growth: A contribution to the problem of scaling in the evolution of size. Amer. Naturalist, 105:113.

-- 1977: Ontogeny and phylogeny. (Belknap) Cambridge.

Grant, F.S. & West, G. F. 1965: Interpretation theory in applied Geophysics. (McGraw Hill) New York.

Grenander, U. 1976: Pattern Synthesis. (Springer) Appl. math Sci. 18, New York.

-- 1978: Pattern Analysis. (Springer) Appl. math. Sci. 24, New York.

-- 1981: Regular structures. (Springer) Appl. Math. Sci. 33, New York.

Güttinger, W. 1979: Catastrophe geometry in physics: a perspective. In: Güttinger & Eikemeier eds.: Structural Stability in Physics. (Springer) Berlin, 23-30.

Güttinger, W., Eikemeier, H. eds. 1979: Structural stability in physics. (Springer) Berlin.

Guggenheimer, H.W. 1977: Differential geometry. (Dover) New York.

Gzovsky, M.V. et al. 1973: Problems of the tectonophysical characteristics of stress deformations, fractures and deformation mechanisms of the earth's crust. Tecto-nophysics, 18:157-205.

Hadeler, K.P. 1974: Mathematik für Biologen. (Springer) Berlin.

Haken, H. 1977: Synergetics. An Introduction. (Springer) Berlin.

-- 1979: Synergetics and a new approach to bifurcation theory. In: Güttinger & Eikemeier: Structural stability in physics. (Springer) Berlin, 31-41.

-- ed. 1981: Chaos and order in nature. (Springer) Berlin.
-- ed. 1982: Evolution of Order and Chaos. (Springer) Berlin.

Hamilton, E.L. 1976: Variations of density and porosity with depth in deep-sea sediments. J. Sediment. Petrol., 46:280-300.

Harbaugh, J.W., Doveton, J.H., Davis, J.C. 1977: Probability methods in oil explora-tion. (J. Wiley), New York.

Harland, W.B. et al. 1982: A geological time scale. (Cambridge Univ. Pr.) Cambridge.
Hartigan, J.A. 1975: Clustering algorithms. (Wiley) New York.

Hattori, I. 1973: Mathematical analysis to discriminate two types of sandstone-shale alternations. Sedimentology, 20:331-346.

Hills, E.S. 1966: Elements of structural geology. (Methuen) London.

Hobbs, B.E. 1971: The analysis of strain in folded layers. Tectonophysics, 11:329–375.

Hochstadt, H. 1964: Differential equations: a modern approach. (Dover Publ.) New York, 1975.

Hoeppener, R. 1978: Grenzen quantitativer Untersuchungen bei tektonischen Experimenten. Geol. Rdsch., 67:858–879.

Honda, H. 1971: Description of the form of trees by the parameters of the tree-like body: Effect of the branching angle and the branch length on the shape of the tree-like body. J. theor. Biol., 31:331–338.

Honda, H. & Fisher, J.B. 1978: Tree branch angle: maximizing effective leaf area. Science, 199:888–890.

— 1979: Ratio of tree branch length: the equitable distribution of leaf clusters on branches. Proc. Natl. Acad. Sci. USA, 76:4875–3879.

Horton, R.E. 1945: Erosional development of streams and their drainage basins: hydrophysical approach to quantitative morphology. Geol. Soc. Am. Bull., 156: 275–370.

Hoschek, J. 1969: Mathematische Grundlagen der Karthographie.(BI) Mannheim.

Hujbreyts, C.J. 1975: Regionalized variables and quantitative analysis of spatial data. In: Davis & McCullagh eds.: Display and analysis of spatial data. (J. Wiley), London, 38–53.

Huxley, J. 1932: Problems of relative growth. (Methuen) London.

Jacobs, O.L.R. 1974: Introduction to control theory. (Clarenton) Oxford.

Jaeger, J.C. 1969: Elasticity, fracture and flow. (Methuen) London, 3rd ed.

Jaeger, J.C. & Cook, N.G.W. 1971: Fundamentals of rock mechanics. (Chapman & Hall) London.

Jänich, K. 1980: Topologie. (Springer) Berlin.

Johnson, D.A. & Damuth, J.E. 1979: Deep thermohaline flow and current-controlled sedimentation in the Amirante Passage: Western Indian Ocean. Marine Geology, 33:1–44.

Johnson, A.M. & Ellen, S.D. 1974: A theory of concentric, kink and sinusoidal folding and of monoclinal flexuring of compressible, elastic multilayers. Tectonophysics, 21:301–339.

Johnson, A.M. & Honea, E. 1975: A theory of concentric, kink, and sinusoidal folding and of monoclinal flexuring of compressible elastic multilayers. Tectonophysics 25:261–280.

Johnson, A.M. & Page, B.M. 1976: A theory of concentric, kink and sinusoidal folding and of monoclinal flexuring of compressible, elastic multilayers, VII. Tectonophysics, 33:97–143.

Johnson, A.M. & Pollard, D.D. 1973: Mechanics of growth of some laccolithic intrusions in the Henry Mountains, Utah. I,II: Tectonophysics, 18:261–309,311–354.

Journel, A.G. & Huijbregts, Ch.J. 1978: Mining Geostatistics. (Acad. Press) London.

Kaesler, R.L. 1966: Quantitative Re-evaluation of ecology and distribution of recent Foraminifera and Ostracoda of Todos Santos Bay, Baja California, Mexico. Univ. Kansas Paleont. Contr. 10.

Kant, R. & Kullmann, J. 1980: Umstellung im Gehäusebau jungpaläozoischer Ammonoideen N. Jb. Geol. Paläont. Mh., 673–685.

Kertz, W. 1969: Einführung in die Geophysik I. (BI) Mannheim, 216–221.

Kléman, M. 1983: Points, Lines and Walls. (J. Wiley) Chichester.

Krause, H.F. 1970: Die von W.B. Kamb vorgeschlagene Berechnung der Auszählfläche für Gefügediagramme, eine kritische Diskussion. Clausth. Tekt. Hefte, 10:175-190.

Krumbein, W.C. 1975: Markov models in the Earth Sciences. In. McCammon, R.B. ed.: Concepts in Geostatistics. (Springer) Berlin, 90-105.

Krumbein, W.C. & Graybill, E.A. 1965: An introduction to statistical models in geology. (McGraw-Hill) New York.

Lindenmayer, A. 1975: Developmental systems and languages in their biological contact. In: Herman, G.T. & Rosenberg, G. eds.: Developmental systems and languages. (North Holland Publ.) Amsterdam.

Ling, F.F. 1973: Surface mechanics. (Wiley) New York.

Lister, G.S., Paterson, M.S., Hobbs, B.E. 1978: The simulation of fabric development in plastic deformation and its application to quarzite: the model. Tectonophysics, 45:107-158.

Lotka, A.J. 1956: Elements of mathematical biology. (Dover) New York.

Love, A.E.H. 1944: A treatise on the mathematical theory of elasticity. (Dover) New York.

Lu, Y.-Ch. 1976: Singularity theory and an introduction to catastrophe theory. (Springer) New York.

Magara, K. 1968: Compaction and migration of fluids in miocene mudstones, Nagavka Plain, Japan. Am. Ass. Petrol. Geol. Bull, 52:2466-2501.

Mandelbrot, B. 1977: Fractals: Form, chance, and dimension. (Freeman), San Francisco

Mangold, H.v., Knopp, K. 1968: Einführung in die Höhere Mathematik. Teil 2. (Hizel Verl.) Stuttgart.

Mardia, K.K. 1972: Statistics of directional data. (Acad. press) New York.

Marsal, D. 1970: Ein Monte Carlo-Test zum Signifikanz-Vergleich beliebiger Gefügediagramme im Schmidt'schen Netz. In: Krückeberg, F. et al. eds.: Computer Einsatz in der Geologie, Clausthaler Tekt. H., 10.

-- 1976: Die numerische Lösung partieller Differentialgleichungen. (BI) Mannheim.

-- 1979: Statistische Methoden für Erdwissenschaftler. (E. Schweizerbart), 2nd ed., Stuttgart.

Marsal, D. & Philipp, W. 1970: Compaction of sediments. A simple mathematical model for calculating the gravitational porosity-depth equilibrium curve of shales. Bull. geol. Inst. Univ. Uppsala, NS, 2:59-66.

Marsden, J.E. & McCracken, M. eds. 1976: The Hopf bifurcation and its applications. (Springer) Appl. math. Sc. 19.

Matthews, P.E., Bond, R.A.B., Van den Berg, J.J. 1971: Analysis and structural implications of a kinematic model of similar folding. Tectonophysics, 12:129-154.

May, R.M. 1974: Biological populations: Stable points, limit cycles, and chaos. Science, 186:645-647.

McGhee, G.R. & Bayer, U. 1985: The local signature of sea-level changes. In: Bayer & Seilacher eds.: Sedimentary and evolutionary cycles. (Springer).

Means, W.D. 1976: Stress and strain -- basic concepts of continuum mechanics for geologists. (Springer) New York.

Meinhard, H. 1984: Models of biological pattern formation. (Academic Pr.) London.

Miall, A.D. 1973: Markov chain analyses applied to an ancient alluvial plain succession. Sedimentology, 20:347-364.

Naef, A. 1928: Die Cephalopoden. Fauna und Flora des Golfes von Neapel. Zool. Stat. Neapel, Mon. 35.

Nicolis, G. & Prigogine I. 1977: Self-organization in nonequilibrium systems. (Wiley) New York.

Niklas, K.J. 1982: Computer simulations of early land plant branching morphologies: canalization of patterns during evolution? Paleobiology, 8:196-210.

Nye, J.F. 1979: Optical caustics and diffraction catastrophes. In: Güttinger & Eikemeier eds.: Structural stability in physics. (Springer) Berlin, 54-60.

-- 1979: Structural stability in evolving flow fields. In: Güttinger & Eikemeier eds.: Structural stability in physics. (Springer) Berlin: 134-140.

Odé, H. 1960: Faulting as a velocity discontinuity in plastic deformation. Geol. Soc. Am. Mem. 79:293-319.

Officer, C.B. 1974: Introduction to theoretical geophysics. (Springer) New York.

Osten, G. & Guckenheimer, J. 1976: Bifurcation phenomena in population models. In Marsden & McCracken eds.: The Hopf bifurcation and its applications. (Springer) Appl. math. Sci. 19:327-353.

Pfaltz, J.L. 1975: Representation of geographic surfaces within a computer. In: Davis & McCullagh eds.: Analysis of spatial data. (Willey) London, 210-230.

Poston, T. & Stewart, I. 1978: Catastrophe theory and its applications. (Pitman) London.

Ramsay, J.G. 1967: Folding and fracturing of rocks. (McGraw-Hill) New York.

Raup, D.M. 1966: Geometric analysis of shell coiling: general problems. J. Paleont., 40:1178-1190.

Raup, D.M. & Michelson, A. 1965: Theoretical morphology of the coiled shell. Science 147:1294-1295.

Raup, D.M. et al. 1973: Stochastic models of phylogeny and the evolution of diversity. J. Geol. 81:525-542.

Rényi, A. 1977: Wahrscheinlichkeitsrechnung. (VEB) Berlin.

Richards, O.W. & Kavanagh, A.J. 1947: The analysis of growing form. In: Essays on growth and form presented to d'Arcy Wentworth Thompson. Oxford, 188-230.

Richtmyer, R.D. 1981: Principles of advanced mathematical physics II. (Springer) New York.

Rollier, L. 1911-1918: Fossiles nouveaux ou peu connus des terrains secondaires. Mém. Soc. paléont. Suisse, XXXVII-XLII.

Rössler, O.E. 1979: Chaos. In: Güttinger & Eikemeier eds.: Structural stability in physics. (Springer) Berlin.

Rosenfeld, A. & Weszka, J.S. 1980: Picture Recognition. In: Fu ed.: Digital pattern recognition. Communication and cybernetic 10 (Springer).

Sampson, R.J. 1975: The SURFACE II graphics system. In: Davis & McCullagh eds.: Analysis of spatial data. (J. Wiley) London.

Schaeffler, D. 1981: General introduction to steady state bifurcation. Lect. Notes Mathematics, 898:13-47, (Springer) New York.

Scheidegger, A.E. 1982: Principles of Geodynamics. (Springer) Berlin, 2nd ed.

Schreider, J.A. 1975: Equality, resemblance, and order. (Mir Publ.) Moscow.

Schumaker, L.L. 1976: Fitting surfaces to scattered data. In: Lorentz, G.G. et al. eds.: Approximation theory II. (Acad. Press) New York, 203-268.

Schwarzacher, W. 1974: Sedimentation models and quantitative stratigraphy. Developm. in sedimentology, 19 (Elsevier) Amsterdam.

Schwarzacher, W. & Fischer, A.G. 1982: Limestone-shale bedding and perturbations of the earth's orbit. In: Einsele & Seilacher eds.: Cyclic and event stratification (Springer) Berlin, 72-95.

Shreve, R.L. 1966: Statistical law of stream numbers. J. Geology, 74:17-37.

Simpson, G.G. 1951: Horses. (Oxford Univ. Pr.) New York.

Smith, R.B. 1975: Unified theory of the onset of folding, boudinage and mullion structure. Geol. Soc. Am. Bull., 86:1601-1609.

Sokal, R.R. & Sneath, P.H.A. 1964: Principles of numerical taxonomy. San Francisco.

Spencer, A.B. & Clabaugh, P.S. 1967: Computer programs for fabric diagrams. Am. J. Sci., 265:166-172.

Stanley, S.M. 1975: A theory of evolution above the species level. Proc. Natl. Acad. Sci., 72:646-650.

-- 1979: Macroevolution: Pattern and Process. (Freeman) San Francisco.

Steinhauser, D. & Langer, K. 1977: Clusteranalysis. (de Gruyter) Berlin.

Stephansson, O. 1974: Stress-induced diffusion during folding. Tectonophysics, 22:233-251

Stewart, I. 1981: Applications of catastrophe theory to the physical sciences. Physica, 2D:245-305.

-- 1982: Catastrophe theory in physics. Rep. Prog. Phys., 45:185-221.

Swift, S.A. 1977: Holocene rates of sediment accumulation in the Panama Basin, eastern equatorial Pacific: pelagic sedimentation and lateral transport. Journ. Geol. 85:301-319.

Terzaghi, K. 1943: Theoretical soil mechanics. (Wiley) New York.

Thom, R. 1970: Topological models in biology. In: Towards a theoretical biology. 3. draft, ed. C.H. Waddington. (Aldine) Chicago, 89-116.

-- 1975: Structural stability and morphogenesis. (Benjamin) Reading.

-- 1979: Towards a revival of natural philosophy. In: Güttinger & Eikemeier eds.: Structural Stability in Physics. (Springer) Berlin.

Thompson, D'Arcy W. 1942: On Growth and Form. (Cambridge Univ. Pr.) Cambridge.

Thompson, J.M.T. 1982: Instabilities and catastrophes in science and engineering. (Wiley).

Trurnit, P. & Amstutz, G. Ch. 1979: Die Bedeutung des Rückstandes von Druck-Lösungsvorgängen für stratigraphische Abfolgen, Wechsellagerung und Lagerstättenbildung. Geolog. Rundschau, 68:1107-1124.

Van Hinte, J.E. 1976: A Jurassic time scale. Amer. Ass. Petrol. Geol. Bull., 60:489-497.

Vail, P.R. et al. 1977: Seismic stratigraphy and global changes of sea level. In: Payton C.E. ed.: Seismic stratigraphy -- applications to hydrocarbon exploration. Am. Assoc. Petr. Geol. Mem., 26:49-212.

Vidal, C. & Pacault, A. 1982: Spatial chemical structures, chemical waves: a review. In: Haken ed.: Evolution of order and chaos. (Springer) Berlin.

Vistelius, A.B. 1976: Mathematical geology and the progress of geological sciences. Journ. Geology, 84:629-651.

Vogel, F. 1975: Probleme und Verfahren der numerischen Klassifikation. (Vanderhoeck& Ruprecht) Göttingen.

Walton, W.R. 1955: Ecology of living benthonic foraminifera, Todos Santos Bay, California. J. Paleont., 29:952-1018.

Westermann, G.E.G. 1966: Covariation and taxonomy of the Jurassic ammonite Sonninia adicra (Waagen). N. Jb. Geol. Paläont. Abh., 124:289-312.

Wirth, N. 1972: Systematisches Programmieren. (Teubner) Stuttgart.

-- 1975: Algorithmen und Datenstrukturen. (Teubner) Stuttgart.

Wright, F.J. 1979: Wavefront dislocations and their analysis using catastrophe theory. In: Güttinger & Eikemeier eds.: Structural stability in physics. (Springer) Berlin.

Wunderlich, W. 1966: Darstellende Geometrie. (BI) Mannheim, 85–104.

Young, J.F. 1975: Einführung in die Informationstheorie. (Oldenbourg) München.

Zienkiwicz, O.C. 1975: Methode der finiten Elemente. (Hauser) Wien.

Zurmühl, R. 1964: Matrizen. (Springer) Berlin.

accumulation 9,92
algorithm 1
allometry 21
alphabets 120
ammonites 19
　　septa 21
　　suture lines 19,143
approximation 40,41,62,79
　159,165
　　bilinear surface 62,67
　　spline 43,67,75
　　surface locally 165
arithmetic mean 31
astronomic cycles 92
attractor 142
autocatalytic systems 142
averaging 31,33,40,41,66,
　137
average
　　moving 30
　　of vector fields 198
　　weighted 49,55
　　iterative 66

bar 207
bedding (planes) 88,90,92,
　95
bed thickness 95
bending 206
Bernoulli's theorem 207
bifurcation 4,70,73,78,108,
　109
　　point 121
　　ratio 116
bilinear function 62,67
binary decision 100
　　tree 98,99,215
biorthogonal grid 155
boundary conditions 42,43
　77,157
box fold 201
branching angle 124,128
　　of trees 114
　　ratio 124
　　solution 214
Buckman's law of covari-
　ance 19,26
bulk density 13

canal surface 149
cardioid 195
catastrophe
　　cusp 50,69,157,168,170
　177,182
　　double cusp 69,189,191

dual cusp 183,187,196
　　fold 181,187
　　hyperbolic umbilic 207
　　stopping potential 177,
　182
　　swallowtail 26,146-
　150,157,172,177,182
　　　modified 174,176
catastrophe manifold 148,
　151
　　potential 170,174,177
　　theory 4,158,160,168,
　183
caustic 79,145,156,167,169
　172-6,180,184,188,193-4
　196,199
centered fan 200
centroid clustering 72,103
　104,105
　　unweighted 104
　　weighted 104
chance 141
chaos 4,70,98,109,112,142
chaotic motion 94
characteristics 201,203
chevron fold 201
chi^2-test 4,72,82
circular reflector 184-6,
　192,195,198
classification 99,102,106,
　144,146,151
　　morphologies 160
　　traveltime record 170,
　171,183
cluster analysis/strategies
　4,98,100,215
　　trees 99,215
　　instability between 106
　　instability within 106
compaction 12,15
complexity, number of
　116
concave reflector 167,
　179,186-7
concentric fold 201
conditional probability 87
conformal mapping 155
consolidation 10,16
constraints 141
continuation of solution
　62,112
contouring 51
contour line 62
convex boundary 55
　　hull 26,215
　　minimal polygons 51,55
　56,57

reflector 186-7
correlation 19
counting area/circle 29-44
critical set 193
critical value graph 174
curvature 112,139,140,150
　155,166,183,186,192,205
cusp catastrophe 56,69,
　157,168,170,177,182,187
cusp point 82,180-181,189
　194
cycles 11,92
cycloid 161,162,
　167,180-1
cylindrical fold 203

D'Arcy Thompson's
　　theorem 132
data
　　directional 29,30,47,72
　82
　　scattered 30,51,
　　sparse 35,54
　　surface 47
deformations 4
density 30
density function/distribu-
　tion 10,31-33,49
　　rectangular 37
depositional models 95,97
difference approximation
　73
difference equation 71,70,
　73
differential equation 71,72
　142
directional data 29,30,47
　72,82
directional search method
　51
discontinuous phenomena
　144
discontinuity, lines of
　201,202
discrete signals 88
discriminant 169
dislocation field 202-3
dislocations 210,212
distance function 58,214
distribution 33
　　uniform 110
disturbed problem 191
double cusp 69,189,191
drainage systems 113,141
dual cusp 183,187-8,196
dynamical system 70

elasticity 198
elementary catastrophe
 cusp 50,69,157,168,
  170,177,182
 fold 181,187
 hyperbolic umbilic 207
 theory 28,144,151,177
  214
 swallowtail 26,146-50,
  157,172,177,182
elliptic reflector 184-5,
 189
 surface point 47
entropy 103,111
envelope 160
equal distance (interval)
 sampling 86,89,92,95
equally spaced grid 36
evolute 150,200,213,214
evolution biological 4,5
 mode 5
 gradual 5
 punctuated 6
evolution of wave fronts
 173,174,177,178
excess hydrostatic pres-
 sure 17
extrapolation 55
extrema, local 35

factor analysis 28
fault 179,199,201
finite net/grid 36
finite point set 4
finite theory of strain
 211,212
first principle of growth
 21
flexure 179
flow in porous media 76
focal line 205
focal surface 213
fold 9,27,179,201,214
 box fold 201
 chevron 201
 concentric 201
 cylindrical 203
 hinges 206
 parallel 201-6
 reconstruction 199-201
 similar 201-6
 trasition parallel-
  similar 205
fold catastrophe 181,187
fold line 144
fold point 188
form, transformation of
 155

Galton's machine 71

Gaussian
 curvature 49
 distribution 24,42
Gauss map 46,47
Gompertz model of
 growth 22
genericity 98,165
geometrical series 127,187
grain contacts 17
grain density 12
grid 113
 biorthogonal 155
 elements 61,65
 equally spaced 36
 finite 36
gridding technique 53
growth
 allometry 21
 first principles 21
 Gompertz model
  22

half-space 201,203
Hencky's theorem 200
Hermite interpolation 43
histogram 31
hodograph of Honda trees
 126,130,131
Honda tree 129,123
Hopf bifurcation 75
Huygens' principle 150,152
 158,163,214
hydrostatic pressure 17
hyperbolic surface point
 47,148,150,189
hyperbolic reflections
 158-9,164-5,171,176,
 179,182,
 189
hyperbolic umbilic 207-8
hyperpolyhedron 106

image concepts 98,100
 inversion 183
 recognition 146
incident rays 192
influence function 30
information theory 8
initial conditions 70,77,
 157
instabilities 4
 within clusters 106
 between clusters 106
interpolation 56
iterated map 70,72,80
intraspecific variability 10
involutes 200

Jacobian
 determinant 193,196,
  199,202,207
 matrix 210-11
Jordan curve 135
Jurassic 11

kink 201

Laplace equation 65
layered media 198,211
leading branch/spiral 131,
 132,139
linear theory of strain
 211-2
lines of discontinuity 201
logarithmic spiral 129,134
 201
loops 121
 singular 72,89,91
logistic function 71,73

Markov chain 86,90
medial axis 151,154
membrane stress 205
minimal convex polygons
 51,55-7
morphology, theoretical
 153
morphospace 4,23,26,214

Navier-Stoke equation 155
nearest neighborhood 106,
 56
nephroid 195
net: see grid
number of complexity 116
numerical taxonomy 98
noise 8
 sampling 41
normal vector 47,150,160,
 197,199,203

ontogeny 10,23,155,157
optimization, singularities
 69
ordering relation 87
orthogonal trajectories
 206
outline of Honda trees
 135,138
overburden 13

parabolic rays 80
 reflector 166,168,173,
 176,179
 182,184,188

228

surface point
47,189
parallel fold 201-6
parallel systems 198,210
partial differential equ.
71
Peano curve 135
periodic function,
sampling 92
phase plane 6,23
phylogeny 5,120,155
picture recognition 146,
151
plane strain 211,212
plasticity 198,200
point of failure 7
polynomial weighting
function 37,40,42,44
porosity 10
porous media, flow 76
power series 127,187
Prandtl's solution (plasticity) 200-1
theorem 200
probability 39,87,114,119
conditional 87
matrix 88,91
transitional 86,87,91
production rule 120
projective plane 148
projection 174,214

random 114,141
rays/ray theory 79,145,159
160,167,181-2,187-9,199
214
incident 192
parabolic 80
reflected 192
receiver 9,80,160
distributed 191
reflection angle 198
hyperbolic 158-9,164-5
171,176,179,182,189
multiple 189
reflector 160,167
circular 184-6,192,195,
198
concave 167,179,186-7
convex 186-7
cycloid 161,162,167,
180-1
elliptic reflector 184-
5, 189
parabolic 166,168,173,
176,179,182,184,188
sinusoidal 162
refinement, stepwise 46
refraction seismology 79
reduction process 123
regular grid 51,53

regular points 144
remote sensing 157
Riemann-Hugoniot
catastrophe 170
ruled surface 58,151

sample size 85
sampling distance 12,94
equal interval 86,89,
92,95
strategies 72
scattered data 30,51
sea-level changes 10
sectorial search method
55
sediments
accumulation 9,10,92
compaction 12,15
consolidation 10,16
bulk density 13
grain density 13
homogeneous 6
stratified 6
thickness 12,15
volume 9
sedimentation rate 92,95
seismic record 4,79,157
seismology, reflection 145
refraction 79
semantic 3
septa 21
s-expressions 120
shadow zone 158,180-1
shape 72
formation 134
shear line 200
Shepard's local method
40-1,52,54
shot point 184
signal 8,86,88,97
discrete 88
similar fold 201-6
similarity 98
similarity index of Honda
trees 129
singularity (theory)
28, 144-5, 214
singular loop 72,89,91
single linkage 112
sinusoidal reflector
162
skeleton of plane figures
_151-2
of trees 117-8
slaved variable 80
slip-line 200,203,214
smoothing error 30
process 29,36
Snell's law 191
sonic velocity 13,176

source location 9,80,163,
191,193
sparse data 35,54
spline interpolation
43,67,75
stability 4
structural 161,198,214
topological 4
'statistical' 35
stereographic projection
29
stopping potential 177-9,
182-3
Strahler number 115
strain 198,210
elements 211
finite 211-2
plane 211-2
stream length 127
stream number 115
stress 198
stress-strain diagrams 6
string of symbols 119-120
St. Vernant's solution
(bending) 207
surface 214
bilinear approx. 62,67
local approx. 165
concave 164
data 47
discontinuities 147
fitting local 40,41
focal 213
fundamental forms
213
ruled 58,151
surface points 47
elliptic 47
hyperbolic 47,148,150,
189
parabolic 47,189
suture line 19,143
swallowtail catastrophe
26,146-150,157,172-4,
177,182-3
modified 174,176
syntax 3

Taylor expansion/series
30,31,166,173,186,194
Terzaghi's model of
consolidation 16
termination point 121
tetrahedron 103
time-delay 18
time scale 5,11
time series 72,86,92
trajectories 6,15,22-3,27-8
214
orthogonal 206

transformation 214
  of form 155
transport equation 76
traveltime record 145,163
  170,172, 174-182,189,
  196
  classification 170-1,183
tree 112,113
  algebraic models 120
  binary 98,99,215
branching 114
  cluster 99,215
  Honda 129,123
  metric model 123

tree-like bodies 72
triangular weighting function 43
triangulation 51,54,58,60,
  61,137
tubular neighborhood 112,
  199

universal unfolding 177
Urysohn's lemma 45

void volume 13,17

water content 13
wavefield 181
wave front 152,158,160-1,
  167,172,174,180,214
  evolution 173,174,177,
  188
  three-dimensional 190
wavelets 158,160,163,194
weighting function 10,40,
  42,46,62
  polynomial 37,40,42,44
  rectangular 42,46
  triangular 43